Hurricane Pioneer

Hurricane Pioneer

Memoirs of Bob Simpson

Robert H. Simpson
with Neal M. Dorst

AMERICAN METEOROLOGICAL SOCIETY

Published by the American Meteorological Society
45 Beacon Street, Boston, Massachusetts 02108

The mission of the American Meteorological Society is to advance the atmospheric and related sciences, technologies, applications, and services for the benefit of society. Founded in 1919, the AMS has a membership of more than 13,000 and represents the premier scientific and professional society serving the atmospheric and related sciences. Additional information regarding society activities and membership can be found at www.ametsoc.org.

Library of Congress Cataloging-in-Publication Data

Simpson, Robert H.
  Hurricane pioneer : memoirs of Bob Simpson / Robert H. Simpson with Neal M. Dorst.
    pages cm
Summary: "The memoirs of Robert H. Simpson, meteorologist, hurricane researcher, and co-creator of the Saffir-Simpson Hurricane Scale"—Provided by publisher.
ISBN 978-1-935704-75-1 (paperback)
1. Simpson, Robert H. 2. Meteorologists--United States--Biography.
3. Tropical meteorology. 4. Hurricanes. I. Dorst, Neal M. II. Title.
  QC858.S565A3 2014
  551.55'2092--dc23
  [B]

                    2014045292

Contents

Robert H. Simpson. (Photo by Darlene Shields)

Preface

This memoir is N O T a comprehensive historical account. The sketches here, chosen from a myriad of other portfolio entries, were selected because they best represent the events that happened to me and influenced the critical decisions I made along the way, changing the course of my 96-year life span—whether for better or worse is the reader's decision—and because they were best for trying to decide how to tell this story and have its purpose understood.

It's been a busy life. But overall a happy, rewarding, and what some (myself included) might consider a successful life. Although most of the way I was much too busy to take account of these outcomes or to inquire why this good fortune—as the years roll on it becomes the more compelling to ask why when I compare what has blessed and fulfilled my life with that of so many of my friends and associates whose talents and capabilities I had always held in higher regard than my own. This account is an attempt to come to grips with the many W H Y s that are worth exploring. While the answers may remain somewhat clouded, it should help explain who I am, or tried to be; but if not in time for me, at age 96, to evaluate it, perhaps it may be for the very few who may discover and find these sketches a curious read.

To my knowledge, neither the Simpson nor Rainey families of my generation produced a historian with professional concerns and motivations to trace and record details of their broader family's history. Nevertheless, both families were concerned with the drama and events that punctuated the lives of each family member within living memory, and enjoyed recalling or repeating stories they had been told by their parents of the encounters or incidents that had made a difference in each relative's life. And what is related here beyond personal encounters has similar sources with the validity of details subject to the frailties of retelling the details. Nevertheless, I am comforted in this regard in the memory that having heard innumerable times the retelling of stories by my immediate relatives, of incidents and notable encounters of family history, the essential details and outcomes remained unchanged.

This memoir covers a life span of 100 years spent in widely varying venues, activities, responsibilities, and opportunities to "make a difference" or to "flub the dub." Like most of my family members and close friends, I succeeded in doing a little of both. I hope to tell the story (God willing) of both outcomes as evenhandedly as my recall of what scientifically shaped my life, and it should be available for the circle of interested family and close friends to review and evaluate as their interests may prompt. I am grateful for the happy and satisfying life I've been fortunate to enjoy, for the opportunities that have come my way to make it so; but most of all, for those whose influence made it possible and for those who were willing to forgive my missteps, most of which I have already forgiven myself, in acknowledgement of my imperfections. The story of 96 years of stimulating, sometimes exciting, encounters is told, as I perceived them—warts and all. They are not told in chronological order because the impacts associated primarily with venues in which they occurred differed in context and sequence from those associated with individuals who greatly influenced my life. So, the first half of the memoir concerns the succession of venues and the nature of their impacts, while the second half is concerned with individuals who helped enrich my life and encouraged my growth potential in one or more ways: by presenting opportunities and motivation, appreciation and encouragement, or understanding and love.

—*Robert H. Simpson*

Chapter 1
My Early Life

Notes on My Ancestry

The Simpsons are a pioneering family. Nearly a century after surviving the Battle of Culloden, they emigrated from Inverness, Scotland, to the New World, crossing Virginia and Tennessee to the Hill Country of Texas before settling in the Marble Falls–Burnet vicinity. My earliest knowledge of these forebears was from stories my father enjoyed telling us of exploration and encounters with the Comanche, American Indian tribe, by my Grandpa Robert Hogan Simpson, a Methodist minister, whose ministry covered the circuit from Austin to Liberty Hill, Marble Falls–Burnet, and back through what is now called Johnson Ranch country to Austin—all covered by horseback. Robert Hogan (locally and in press references known as "Bob") was married to Margaret Elizabeth Moore (whose ancestry, I am told, has been traced back to the fourteenth century by a retired U.S. Army Col. Moore of Fairfax, Virginia). I never became acquainted with my maternal grandmother's ancestry or their involvements. Strangely, we children rarely overheard accounts or stories about encounters of family members of either Margaret Elizabeth Simpson or Margaret Elizabeth Rainey, who both bore the same Christian names (although Margaret Elizabeth Rainey was informally known as "Maggie").

Robert Hogan and Margaret's children were Daisy of San Angelo, Texas, the eldest; Albert Dee, prominent banker and protégé of Jesse Holman Jones of Houston; Katherine (Kate), of Corpus Christi; Stanton Fields, hardware executive of Corpus Christi, and Clyde Robert (my father), a ranching equipment merchant of Corpus Christi.

The Rainey family, in my mother's generation, was widely known in western Tennessee, principally in Brownsville, but with roots in earlier generations in Haywood County. At home in Corpus Christi, we children rarely heard the conversation turn to earlier generations, and questions asked of my earlier forebears were usually answered curtly and incompletely. Grandma Rainey (Maggie), with several marriages, rarely referred to or discussed her parentage or earlier ancestry, which essentially remained a blank page in my Simpson family circle. Nevertheless, the Tennessee Raineys of my mother's generation were a tightly bonded, caring family and highly regarded and respected leaders in the Brownsville community. But I cannot recall a single occasion when our Grandpa Rainey's name entered the conversation at our dinner table. That said, a brief résumé of the Rainey ancestry was compiled but selectively distributed by Donis Wolfe, son of Elizabeth Rainey Wolfe (stepsister of my mother). Maggie Rainey had a total of eight children, many of whom died at birth or in early youth. The surviving children were Roy, a grocer and cotton broker of Brownsville; Annie Laurie of Corpus Christi (my mother); Clyde, a printer in Corpus Christi; and Homer, a banker in Brownsville, Tennessee.

My Childhood

These were the family roots from which I arrived in Corpus Christi, Texas, on November 19, 1912, all 8½ pounds of me, son of a beautiful 23-year-old mother and a handsome 25-year-old intellectual-minded father—both very active members of the First Methodist Church and in community affairs. Clyde Simpson, my dad, was a successful and ambitious proprietor of a ranching hardware establishment. Annie Laurie Simpson, my mom, was a pianist and was employed by the elegant Gunst Piano Company to demonstrate Steinway pianos and new releases of sheet music. She was also active in a number of societies, including the Eastern Star and Women's Temperance Union.

My parents' first residence in Corpus was at 1219 Chaparral Street, a little north of the main business area, 2½ blocks from our church, 3 blocks from the courthouse, and several more to the center of town. Ours was a sturdily built eight-room, two-story home; spacious and comfortable, but not as elegant as the numerous other residences within a short distance of us in a fashionable area of downtown Corpus—fashionable yes, but sociable NO. The Simpsons were not poor and enjoyed a comfortable lifestyle, but they were not wealthy enough "to keep up with the Joneses," in the common parlance of society in Corpus, a friendly small city without odious social prejudices. But as frequently is the case, wealth, not common friendliness, tends to prevail in the cultivation of close social friendships.

Also within two blocks or so of 1219 Chaparral, there was never a child near my age that could become my playmate. While I tend to doubt this environment was a major factor, it was soon evident that I enjoyed aloneness and "doing my own thing," rarely showing signs of loneliness. Clearly I loved and depended on my family; but equally clear, the family was not a primary focus of my interest and attention. I rarely played indoors, and occasionally I snuck away from home on my tiny tricycle to explore unfamiliar areas of the town. It was then I became an explorer (at heart), much to the dismay of my mother and the enthusiastic acceptance of my dad. Because of this, my dad would forever be my hero, while my mom would have to make do with my love and respect. Not that I was dissatisfied with my home life or the concerns and affection of my parents, but I was captured by the urge to know more about the world around me and what it had to offer.

Our house was also little more than a block from the water's edge and beach of Corpus Christi Bay, which had a very small tidal variation. As I later learned, my dad was eager to reside in the downtown area and near the water but was equally concerned to acquire a property that would withstand not only hurricane force winds but, within reason, the threat of storm surge (or "tidal waves," as it was known in those days). In selecting 1219 Chaparral, he felt he had found the right solution, and almost did, except for the storm of September 14, 1919, discussed later. Either an engineer or a very good architect must have designed this house: the foundation consisted of reinforced concrete walls three or four feet above the surrounding terrain in the shape of a cross to which the sturdy frame of the house was bolted; the foundation providing a living room floor level at least six feet above grade level and some eight feet above mean sea level.

It was not until my college years, however, that I learned Dad's greater fears in this procurement. The acquisition of this house had hinged on his agreement to accept funds provided by my Grandma Rainey (Mubba, as her grandchildren knew her); in exchange for the money, Mom and Dad would care for all of Mubba's needs for the rest of her life. They did so but at much more than the monetary cost. She remained a housebound cripple in a wheelchair for her 31 remaining years—at least so in the later perception of his children. But without open complaint, they shouldered it; although in later years, it showed in Dad's face.

Our home on Chaparral was spacious, but it was fully occupied with both family and renters. Besides Mom, Dad, Mubba, and me, there were my Uncle Homer Rainey and Aunt Daisy Simpson, and the middle-aged Hensley couple, who rented an upstairs bedroom. In addition there was an occasional Rainey family visitor from San Antonio, Texas, or Brownsville, Tennessee. Our Corpus relatives, Uncle Stant and Aunt Carra Simpson, who lived in a comfortable two-story residence on Carrizo Street several blocks inland from the bluff, near South Bluff Park, often joined us for dinner and a game of "forty-two" (a bridge-style game played with dominoes, popular in Texas). Stant was a traveling salesperson with Corpus Christi Hardware, a wholesale outlet. Dad established his own hardware business, supplying heavy ranching equipment, tractors, and GMC trucks; his principal client being the King Ranch.

Corpus, a "big" little town of 12,500 when I first made my appearance, even then was a sophisticated and proud community in a picturesque half-moon bay. Its business area and many of the more elegant residences spread gracefully inland three or four city blocks to where a bluff rose abruptly 30–40 feet to Broadway Boulevard with its well-spaced elegant residences and a few mansions affording a magnificent view of the town below. Its waterfront and recreational centers included several wharves for fishing vessels and, best of all, dockage for the venerable *Japonica*, an 80-foot, broad-beamed motor cruise vessel, best known for its memorable moonlight cruises on Corpus Christi Bay. To the discerning visitor, Corpus was unique not only for its magnificent setting along the coastal bend of Texas, but for its distinctive architecture that graced the downtown area—perhaps the better examples being the six-story Nueces Hotel, the most elegant accommodation in Texas south of Houston and north of San Antonio; the five-story county courthouse; and especially the First Methodist Church, a tall domed structure with a circular sanctuary, but without a steeple. In those years it was the largest church in Corpus, with the most distinctive church architecture in south Texas.

Below the bluff, the city layout was strung out in a north–south line to either side of two principal adjacent streets: Chaparral (the principal business street) and Mesquite Street, each extended southward about a mile and a half from what is now the ship channel and port area to what is known as South Bluff. Corpus was proud of its distinctive streetcar system, which extended from its northern terminus down Chaparral to a merger with Mesquite and then westward up the sides of a broad arroyo to South Bluff and its less sophisticated South Bluff suburbs.

These details of the geographical scope of my youthful wanderings will prove their worth in episodes described in later sketches; for example, in the early 1920s, a few young pranksters, whose families were acquaintances of Mom and Dad, celebrated Halloween by greasing the tracks of the streetcar on the incline leading up to South Bluff following the right turn westward from Mesquite, hoping to enjoy the futile attempt of the streetcar to climb up the incline to South Bluff. Unfortunately, the lark turned to disaster and disgrace when the first streetcar to arrive was at the top of the incline. The accelerating vehicle had no difficulty negotiating the hill but jumped the tracks when it failed to make the turn at the bottom and crashed through a furniture store with an embarrassing amount of damage to the streetcar, the store, and its contents—not to mention the disgrace and despair suffered by the pranksters, who spent most of the night at the police station before being retrieved by their parents. My dad at our lunch table the next day effectively delivered the object lesson, a rude but memorable example of the need to "think of the consequences before you decide on your prank!"

All this was the Corpus Christi that nurtured my childhood and pre-school life, captured my interests, and fueled my earliest ambitions (which were numerous). But there was a notable missing link—whether for better or worse, it will likely remain unresolved. There were no children or playmates within blocks of our home anywhere near my age with whom I could bond, share, or play kiddies' games—no one to compete with; no one to hone my talents, creative skills, or leadership proclivities. If I became a loner as a result, it never resulted in loneliness. But it did teach me, early on, that there's a wondrous world out there, but it's not going to seek you out; if it interests you and you are willing to face its challenges, hop to it.

If 1219 Chaparral offered little to attract pre-school children, then the city maintained a wide variety of social activities for adults of all ages in which my family participated. My mother was active in the Eastern Star chapter and in the Women's Temperance Union; Dad, in the Kiwanis club. Families gathered in City Park each Sunday afternoon for a band concert in the elegantly designed bandstand. But our family's principal social involvement was in multiple activities at the curiously domed First Methodist Church, with its fine organ and adjacent Epworth League building. Easter and Christmas pageants were always memorable. Both Dad and Uncle Stant were members of the Board of Stewards; Dad, with his fine tenor voice, sang in the choir and with the church quartet. Both parents had leadership roles in other routine church functions and during twice-annual special services, led by visiting well-known evangelists. I remember particularly the visits of Homer Rodeheaver, who with his gold trombone led the songfests, but even at my young age, I was (at best) impatient, if not perturbed, by the athletic-style antics from the pulpit by some of the evangelists.

Not until Mom and Dad bundled me into our 1915 Ford "touring car" together with a family picnic lunch and drove the three of us the long 150 miles through such villages as Taft, Sinton, Beeville, and Kennedy, including the rain-drenched muddy dirt roads of Wilson County, to San Antonio did I find reason to appreciate how fortunate we were to live in Corpus Christi. Though only four years old, I well remember the impact of this journey, as completely as I occasionally recall now, more than nine decades later, when my mom, in my earliest years, held me in her arms in her favorite rocking chair, gently singing a memorable Charles Wesley hymn until I fell asleep.

The Impact of My Family

There is no question that these early experiences set the tone—if not the rapid pace—of activities that filled a life of adventure and exploration: a quest for explanations or understanding of many puzzlements and processes that others close by were quite satisfied to accept by faith. In my preteen years, family life at home was always a venue for giving vent to curiosity, asking questions, and seeking answers. At times it was as simple as understanding the meaning of words; in which case, the answers were easy: the dictionary was the authority, and on one side of the dinner table, or nearby, one was always available. For nearly everything else, the Bible was at Dad's side and he was pretty skilled at finding enlightenment of a kind there. But when nagging questions proved a block to understanding, puzzlement was replaced, cognitively, by faith: a term whose definition, conversationally, was accorded a broader meaning than the biblical one: that is, accepted as truth, in search of a fuller understanding (explained, of course, as necessary in language appropriate to the child's age; e.g., "lots of things we know must be true are hard to understand or explain completely until we get a bit older and wiser"). At least for the Simpson children, in this context the problems that some encounter in the aura and acceptance of the mystical as a concept were not only alleviated, but the thirst for knowledge and understanding was stimulated. Only years later, during my lone and sometimes lonely hours, in the absence of a family of my own, trying to piece together how and why I had been so fortunate as to end up in Hawaii with the opportunities confronting me there, did I realize how unique and effective a childhood contribution my parents had made in my ability to cope with what life had dished out, not only with the challenges and opportunities, but with the strength and determination to cope and rise above the failures and questionable decisions I had made along the way.

Annie Laurie (My Mom)

There must have been many, but I can't recall a mother who openly displayed the affection and pride in her first born, and the ambition to see him have every opportunity to succeed, as my mom held for me. But while her efforts in this regard were overt, in retrospect I never detected any evidence of selfish motivations, nor desire to make me into a mama's boy, nor mold me in her own image or social preferences, so long as I displayed no signs of becoming a maverick in regard to religious and social conventions. Certainly her love and efforts on my behalf did not escape my attention, and I hope my appreciation was as overt as her efforts.

Unfortunately, this proved to be a struggle for me, mainly because the opportunities and suggestions she continually thrust in my path, though never mandated, were rarely attractive or consistent with the preferred lifestyle of independence and social comfort as a loner, though by no means that of a hermit. Travel, mainly in search of understanding, was my fondest cup of tea. But Mom was quick to recognize my love of music, stirring religious music, and my ability to sing at an early age and to tap my right foot in time with marching band music.

Mom lived a people-centric, extrovert's life and was a marvelous cook, always planning picnics and nearby beach parties. She loved having my college friends, who frequently accompanied me home from Southwestern University, as overnight guests, to whom she was forever known as "Mom Simpson." Having a weekend guest in the Simpson household was invariably a memorable event, both for the guest and for me. But to spend Sunday afternoon or a weekend exploring the fascinating countryside, or visiting close friends or relatives in San Antonio or elsewhere upstate was not her cup of tea, although much preferred by my dad and me.

Clyde (My Dad)

In no way could it be regarded as a rejection of my mom that my dad was always my intensely admired hero. He was not particularly extroverted, but he loved sports. Travel, however, was more enticing to him, and exploration, not visitation, was its primary motivation. He was an intellectual, extensively read, a knowledgeable historian with active interests in political issues (but not in the machinery of politics per se). In short, I adored him as a father, admired him as a human being, and we enjoyed each other, sharing opinions and fascinating questions.

If Dad was widely regarded as a devoted family-centered man and devout Christian, then he was equally well respected in the business community, not only as a successful merchant, but also as an icon of integrity and ethical business practices. He was a community leader and a member of the Kiwanis and other charitable business organizations. Unfortunately, in later life, his leadership in community affairs was handicapped by his lingering battle with pernicious anemia.

The Hurricane of 1919 in Corpus Christi

On Sunday morning of September 14, 1919, we awoke early, with the windows at our 1219 Chaparral residence in Corpus rattling, and the palm fronds out front lurching singularly southward, parallel to the bay shore, which was just a block east of Chaparral Street.

My dad and Uncle Homer left the breakfast table early with a look of uneasiness on their faces and ventured out into the heavy rain to see what was going on at the bay front. They returned very shortly, reporting that the bay was in turmoil and waves were breaking inland from the usual shoreline. The flooding extended inland more than half a block. Dad commented, "Well, at least we'll not be going to church today." Mom looked up in surprise but continued preparing the usual celebratory Sunday dinner. Suddenly, we heard the siren of a fire truck making it way slowly northward down Chaparral, splashing through the shallow salt water then covering the street. From a bullhorn on the fire truck, the residents were being advised that a hurricane was approaching and that safer shelter should be sought on higher ground.

The Hensleys, our boarders on the second floor, were spending the weekend in San Antonio. But the Simpsons and Raineys, five in all, remained in harm's way. When Mom announced that dinner was ready, my worried Dad, noting that the rising water had reached the steps from our sidewalk to our front porch (at least three feet higher than it had in 1916, when a similar hurricane had passed over Corpus Christi) stated in voice that brooked no questions or discussion, "It's too late! We're leaving, NOW!" And we did—Grandma Rainey, strapped into her wooden cane wheelchair towed by Uncle Homer; me, on Dad's back; and my mom, bringing up the rear with a double brown bag filled with hot fried chicken fresh from the skillet and some fresh doughnuts, all salvaged from the dinner table at the last moment. The refuge Dad had decided to attempt to reach was three blocks away in the well-built multistory courthouse structure located on a high grassy terrace. The route Dad had chosen carried us over a backyard fence and alleyway to Mesquite Street and then southward two blocks to our hoped-for refuge. With the wind at our backs, wading and swimming, through chest-high water, we somehow made it despite two sources of trauma: First, and to me the worst of the two, the loss of the aborted dinner—as Mom's arm that was carrying the sack of survival vitals tired, the bottom fell out of the wetted sack and with it the contents, the doughnuts floating past us in the strong winds; second was the blood on our faces when we entered our refuge with cuts from windborne beach sand. We weren't exposed to the strongest winds while en route to the courthouse, but records reported the highest winds measured at the weather station about four blocks away before the anemometer was blown away at about 110 miles an hour.

Once we were safely relocated on the sixth floor of the courthouse, looking down on the fury raging below, we were aghast at what had already occurred and was still occurring. Not only was there vast wreckage everywhere but houses, still intact, were afloat, many with refugees clinging to them. My saddest memory of that awful melee was of a sizable cistern on its side, rotating as it moved southward down Mesquite Street, with a man trying to maintain his perch atop it while carrying a baby in his arms. Incredibly, he managed to keep his stance atop the cistern until it reached the first intersection, where the swirling waters caused him to lose footing, plunging him and the baby into the flooding storm surge below. After an incredible period of submersion, his head reappeared but without the baby! Seemingly without hesitation he dove back under, but he was never seen again by the tense onlookers from their sixth-floor refuge. What a dreadful experience to be thrust on a 6 ½-year-old lad who never ever forgot that had it been Friday instead of Sunday that the unexpected hurricane had struck, he undoubtedly would have been at David Hirsh School on his third, and probably last, week of public education. David Hirsh School was a total loss that fatal day.

Early the next day, Dad joined his close friend, David Peel, Corpus's principal undertaker, and a party of volunteers to search for survivors and to recover the bodies of the many who didn't. To this day the total number of fatalities remains uncertain, ranging from early reports of about 250 to later estimates of 500 or more, owing mainly to the uncertain numbers of tourists and nonresident industrial fisherman unaccounted for. The downtown business area below the bluff was close to a total loss, recovery requiring almost two decades.

Meanwhile, my mother and I ventured forth on Monday after the hurricane to see what was left of our residence, and if it was habitable. It wasn't! The surge waters that swept across Corpus Christi Bay first destroyed large tanks of petroleum products near Port Aransas. This was swept westward across the bay, gathering splintered debris as it went, and depositing the debris on both exterior surfaces and, to a lesser degree, interior areas of dwellings and other structures in Corpus as the surge subsided. This left a noxious mess, difficult to clean, making habitability difficult to restore. When Dad first set eyes on its impact at 1219 Chaparral, he turned away, muttering, "We'll never set foot in this place again." However, what Mom and I saw inside the house when we finally gained entrance to the smelly interior was curious beyond belief—dishevelment with minimal destruction: for example, a china closet in the dining room, doors wide open, with unbroken glass from which a stack of 12 china dishes had been removed and relocated, unbroken, on the back of the upright piano, which lay face down in the living room, all without so much as a chip removed from a dish. But Dad had his way. After removing a number of salvageable items, we never again entered that unsightly smelly residence. After several disruptive moves to a various rental properties, Dad purchased a substantial residence at 1223 Second Street, a block inland from the 40-foot bluff at the bay front south of the city.

Hurricane Recovery

1223 Second Street was in an established neighborhood of South Bluff, although still accessed by a dirt road, and several miles south of what, in the eyes and memories of most prehurricane residents, had been, in effect, the almost totally destroyed business center of Corpus. In the four years following the hurricane, most of the uglier scars of the hurricane had been cleared away. But there was precious little venture capital eager to rebuild the low-lying business area below the bluff, and entrepreneurs were cautiously confining their attentions to residential properties in a new subdivision in the South Bluff area known as Del Mar, farther south and west of the old business area. Moreover, merchants, having suffered enormous, mostly uninsured, losses in the hurricane, were loath to risk reestablishing their businesses below the bluff or among the ragtag hovels west of the bluff. Fortunately Dad's business, though only a block from the water, facing away from the onslaughts of wind and waves, survived intact with relatively small losses of inventory. It was open again for business within a few weeks, as was the similarly exposed J. C. Blacknall Dodge Agency across Starr Street, although the Buick Agency immediately across Chaparral facing the bay front was not so fortunate.

1223 Second Street, a quite different venue from that of my seven years at 1219 Chaparral. It was close to quite a few children, especially from County Judge Wright's large family, whose children shared a number of athletic activities with me that Dad acquired for me, including gym equipment and a several-hundred-foot lot cleared for sandlot baseball. However, these activities, shared with neighborhood friends, soon gave way to a new and solitary interest: music. My mom, who as a piano teacher had failed to attract me as her student, persuaded Dad to buy me a new Holton trumpet, with which I promptly fell in love. And I was fortunate enough to become the student of George Boiteau, a former principal clarinetist with the Boston Symphony, who had retired to Corpus for health seasons. And even more fortunate, a few years later, I became associated with Arthur Harris, a recent graduate of the Cincinnati Conservatory of Music. This diversion was destined to have a significant influence in paving the way for my later career in science.

Clyde Roy and Daisy May (My Brother and Sister)

Clyde Roy, my brother, was not born until a year after the 1919 hurricane, nearly eight years younger than I. Daisy May, my sister, didn't make her debut until nearly three years later, in 1923. Both spent their childhood years during my adolescence and college years, most of that time while I was away from home and unable to share my childhood experiences with them. Worse yet (some might contend, better), when I did return home, it wasn't as a brother, but as a high school teacher and band and orchestra director in Corpus Christi High School.

Early Experiences in Driving

The police department in Corpus certified me to drive. There were no such things as a driver's license then—you were simply placed on a police registry of qualification to drive. All that was required was an affidavit by my dad that I was a competent driver. He took me to the police station to meet the chief and then left the office after witnessing the entry of my name on the registry. Competent??? They had Dad's word for it, and Dad was well known around the city. How could Dad be so sure? I was only 11 years old at that time: it was not a common occurrence, but a credible one to all who knew Dad well.

My driving experience began this way: for nearly two years, one of my favorite pastimes, aside from my interests in music, was accompanying my dad on his business travels across the countryside to market steam tractors, heavy farming and ranching equipment, and GM high bed trucks for which Clyde Simpson Hardware was a principal purveyor and source. On one of these trips, Dad left me in his Model T Ford for a few minutes while he tracked down a client at work several acres off the road. I welcomed the opportunity to experiment with the Model T, which I had carefully studied as Dad operated it en route there. The results were not quite what I had expected, but sent Dad storming back to find out what he had apparently anticipated. If he was angry when he arrived, it didn't show outwardly. But his voice had a sharp edge to it when he stared me in the eye and said (as I now remember), "Well, if you're that anxious to learn, let's learn it right." So, for the next half hour or so with no traffic on that country road, I got my first supervised lesson in driving a sedan, attentively, if sheepishly. And before long, I was operating the small tractors and trucks to their display positions in Dad's store.

The culmination of these exciting experiences came in the early summer of 1928. Dad had just sold a large GM truck that begged delivery within a week and was not in his inventory, the nearest source being a distribution center in Dallas, Texas. At breakfast Dad startled me by asking if I thought I could drive the truck from Dallas to Corpus for him. After gulping a couple of times, I replied, "Yes, if you think I can do it," but I reminded him that I had never driven a car or truck outside the Corpus city limits, had never visited Dallas, and had no idea of the road system to make my way back to Corpus. He, of course, knew that but explained that while road maps across Texas were not commonly available in those days, he had already called the distributor in Dallas and arranged to have him sketch out the road numbers and markers I should follow along the route from Dallas to Waco, Temple, Austin, San Antonio, and home to Corpus after spending the night with the Rainey family in San Antonio. Dad put me on the train in an exciting Pullman berth that night—my first such experience. But I didn't get much sleep while contemplating the uncertainties I envisaged ahead. When I arrived in Dallas, a short taxi ride put me in touch with the distributor, where I exchanged the necessary paperwork, and found my way on the road southward toward San Antonio by midmorning, and reached there shortly before dusk, stopping only for fuel and a snack for lunch. Frankly, I can't recall now, once in San Antonio, how I found my way back out to Fredericksburg Road to its distant suburbs and to the Rainey homestead all without incident; but I did. There they were, worriedly expecting me. The next morning they guided me across town to the southern suburbs en route to Corpus. I arrived in midafternoon to deliver the GM truck to Dad's downtown hardware store without a blemish. I was very tense and weary but proud of how easy this challenge had turned out to be.

The Mississippi Flood of 1927 and Our Vacation Travel to Rainey Land

The Clyde Simpson and Clyde Rainey families decided to travel together on a vacation trip to visit Mom's remaining family in Brownsville, Tennessee, in the summer of 1927. The Simpsons—Mom and Dad; Clyde Roy (age 6); Daisy May, also known as Mayme (age 4); and me (age 14)—were in Dad's new disc wheel Dodge sedan, and the two Raineys, Clyde and Edna, were in their not-so-new Nash. Their son, Leroy, was attending summer school and couldn't accompany us. It was not a propitious time for road travel on this route, at least for the worried adults. But for the kids, it was a memorable lark making one detour after another. Having traveled this route several times earlier, Dad was not concerned about the Mississippi River flooding from Oklahoma to Memphis, Tennessee, remembering the sturdy bridge and its lengthy raised approaches to the main bridge to Memphis. But what we didn't learn in time was the damage done along the tributaries, especially by the White River, where bridge after bridge had been destroyed or was inaccessible because of flooding of the approaches.

At this point my memory of the sequence of strange but fascinating sites and occurrences are now confusing to distinguish, although when later I tried to reconstruct it all, I was reminded of portions I had read earlier from Kipling's *Jungle Book*, though I had never experienced travel of any kind through a jungle.

However, when we reached the White River in Arkansas, we were advised that the only way "within a hundred miles" we could cross it was to board a makeshift barge with our cars, and ferry about 10 miles downstream, and off board on a road from which we could reach the Mississippi bridge. This proved to be the fun part for us kids. But for the adults, only a period of anxiety, sufficiently so that none of us worried that we had had little to eat since the evening before. The river wound through a heavily wooded area with a fascinating abundance of birds, snakes, and other wildlife, but was largely unpopulated. This travel had coincided with peak flooding of the Mississippi and its tributaries in Tennessee, Oklahoma, eastern Texas, and Mississippi, where highway bridges had been washed away or were impassable because of flooding on approaches to the bridges.

Once the ferrying was complete and we all looked forward to crossing over the high bridge into Tennessee, hunger pangs reminded us how long it had been since we last ate, only to discover that neither of the cars could climb the river banks, muddied by rains the preceding night. So, we had to wait till a team of mules could be fetched to tow our cars to a level roadway before continuing to the nearest general store to assuage our hunger.

What a wonderful sight Brownsville was when we pulled in near dusk. And to add a welcome mat for the weary travelers, Uncle Roy, having received Dad's phone call from Memphis, had arranged for his housekeeper and cooks to stay late and spread his traditional banquet for us. As tired as we all were, this could not have been a more welcome howdy after a brief opportunity to freshen up.

Despite the memorable and extraordinary experiences of our trip to Brownsville, I was to be treated to two more unique experiences (unique at least for me) before returning home. At breakfast next morning, Uncle Roy asked me to come visit him at his grocery store (it turned out, he was more proud of this proprietorship in Brownsville than of his business in Memphis as a cotton broker and assessor). His sons, Robert and Bill Rainey, advised that they had arranged a blind date for me to accompany them to a local dance that evening. Again, it turned out that both invitations were more of a first for me than I could have imagined earlier.

Uncle Roy proudly showed me around Rainey Grocery and explained why it was so special, and then invited me up to his balcony office, where he extracted a quart bottle of strawberry liqueur, saying, "You'll enjoy this; it's the best in this state." I downed the double thimbleful cautiously, managing not to cough or say what I was thinking: "My dad would not be proud of me for this, and Mom would be furious!" It was the first alcoholic beverage I had ever tasted; I didn't like it, and I was curious why I should feel ashamed.

That evening, quakefully and filled with uncertainty, I attended my first public dance, although I had had a few lessons at the Episcopal church in Corpus to which I was occasionally invited as a chaperoned Friday night social affair at the church. I had confided in Bill in advance concerning my lack of dating experience and, bless his heart, he understood and did everything possible to make me feel comfortable; and I was, after a fashion. I wasn't fond of my date and remember nothing about her. And I was sure she felt the same about me, although it seemed obvious it wasn't her first date. Oh, well, there's a first time for everything, I mused. I was not disappointed, either in my unfortunate date or myself.

I graduated from Corpus Christi High School in 1929. The graduation exercises and my class's final production of *Romeo and Juliet* were held in the new auditorium on its spacious, well-equipped stage. Both events proved memorable for quite different reasons. My performance in the role of Capulet made it clear to me I was not cut out for a theatrical career. But my assigned role in the graduation services, a trumpet solo performance of Frank Simon's "Willow Echoes," went much better…save for one incident (embarrassing for me, but mirthful for the audience). The second section of "Willow Echoes" (otherwise, a beautifully lyrical piece) involved 24 bars of rapid triple tonguing, ordinarily not a difficult passage but on this occasion, at best, memorable. I was wearing, for the first time, a new dark blue surge suit, white shirt with a stiff collar, and a prettied black bow tie that rested comfortably astride my Adam's apple. This was no problem at all until I entered the triple tonguing passage that set my Adam's apple fluttering up and down, and with it the more visible black bow tie in an athletic-type rhythm. It didn't go unnoticed by the audience; as for me, as later reported, I had a very red and strained look on my face. To cap it all, Uncle Homer and the other Raineys, delayed by having to get through the muddy roads of Wilson County to attend the graduation services, arrived just in time to observe this light note injected into an otherwise solemn ceremony. All of which was a preamble to an unanticipated sequence of events that occurred in the following six months that drastically changed the remainder of my life.

Chapter 2
My College Years

My Road to a Career in Architecture

The public school system in Corpus Christi was ranked among the best in the state, during the latter 1920s and my high school days. But it succeeded in attracting my better efforts only in mathematics, English, music, and architectural drawing, a curriculum oddity promoted by a Mr. Pinson, a talented and eager engineering drawing teacher, who had had a difficult time earning a living in architecture. He did succeed in convincing me I had the talent to become a successful architect, and managed to get me a summer job in the county engineer's office in 1928, drafting a revision of the Nueces County road systems. After graduation in May of 1929, Uncle Clyde Rainey from San Antonio, who had been aware of my search for a job, said he might know of career opportunities for me. He contacted Dad to tell him of an ad in the *San Antonio Express* for qualified applicants to design houses for a new subdivision of upscale homes out on Fredericksburg Road, near where the Raineys lived. Dad and I drove to San Antonio, carrying a letter of recommendation from Mr. Pinson; met and interviewed first with a Mr. Buchek, field manager of the new development; then with the developer himself, Mr. L. E. Fite. With little delay, I was employed at (as I now seem to recall) $75.00 a week as an apprentice architect, working on the development site with a junior architect. It was a pleasant and welcomed job. Our task was to design upscale residences whose market cost ranged from $10,000 to $12,000 and architectural design ranging from Spanish to European, none adjacent with identical or similar style or design. Mr. Buchek reviewed our work weekly, suggesting various changes or approving the design for construction.

It was truly an exciting as well as challenging experience, one that could well have carried me to a blossoming career in architecture. If all went well, Mr. Fite proposed assistance in my obtaining a scholarship toward a degree in architecture at the University of Texas. But it didn't! The onset of the Great Depression just a few months later saw to that. Before Christmas my job, and with it my enthusiasm for a career in architecture, came to a rude end. It didn't take much convincing to decide I should seek a different career, one that carried me to Southwestern University in Georgetown, Texas, and one that led to my career in science, which made it possible and supported by my interest and skill in music.

A Christmas Gift

An unexpected present arrived through a telephone message at 1223 Second Street a few days before Christmas 1929. It was from a man I had never met personally but recognized by name. He was one of the judges at the annual state music contest at the Texas College of Arts and Industries in Kingsville. He introduced himself as Edward P. Onstot and recalled that I had participated in the instrumental solo contest, winning a first division rating in the trumpet solo two years in a row when he had been a judge. From somewhere he had learned that I had just lost the job in San Antonio and wondered if I would be interested in coming to Southwestern University and play in his band and the small orchestra that he led there. He quickly added that he knew everyone was having financial difficulty with college tuitions, but that if I wished to join him he would see I got sufficient opportunity to work my way through to a bachelor's degree. However, if my answer were yes, I would have to report to the campus by the second week of January. I tried to mask my exuberance, yet in my reply I'm sure it seeped through. Anyway, Dad borrowed the $150 from a friend, J. B. Jones. We felt that was the least amount of money with which I should start college. I not only showed up in Georgetown on time, but I managed to work my way through the completion of a master's degree in physics at Emory University in Atlanta in 1935 without further borrowing, although I did not get around to repaying the first $150 until years later.

Southwestern University

As a youthful student in Corpus Christi, my academic record was acceptable but little more. But from the day I matriculated at Southwestern University (SU), the first week of January 1930, my life turned a corner. This brought a broadening of my interests and carried opportunities to within my grasp. The teaching staff was more concerned with steering me along a path that would make the best use of my potential than in cramming the contents of textbooks into my memory. It was indeed a glorious thing to feel both capable and appreciated. And after my second scholastic quarter, as best I can remember, I never scored less than an A in coursework, and by my second year at SU, I was elected a member of Scholarships of the South. My sense of self-fulfillment and enjoyment of life grew by the month at Southwestern, and my tenure there was filled with unique experiences and memorable encounters.

Southwestern proved to be a unique experience for many of my friends as well—not just that it is the oldest university (actually a liberal arts college) in the state of Texas, founded in 1840. Struggling to keep its head above water during the Great Depression, Southwestern retained its proud faculty by paying them a (barely) living wage in script accepted at face value by Georgetown merchants. The faculty was determined to maintain SU's tradition of graduating well-educated students with the motivation and self-confidence to become successful leaders in their chosen professions. And a significant percentage of its (then) small student body became nationally acclaimed leaders in medicine, the physical sciences, religion, and music.

Many students in those days, as I did, arrived in Georgetown virtually penniless, being attracted by promises of enough work to pay for most of their early years through college. SU at that time owned a multiacre farm adjacent to the campus that supplied much of the food needed for the elegantly administered dining hall while also supplying employment for many students. I entered SU with that $150 loan from J. B. Jones. This, together with income earned from multiple jobs I worked at, and the few scholarships I acquired, sustained me for the next two years. Working my way through SU during the worst of the Depression years was more of an adventure than a challenge, or worse, a Depression-related hardship. I soon learned the rewards of independence, fiscal and otherwise—that I had the ability to accomplish almost anything I set out to do if I wanted to work hard enough. And with it I soon learned not to worry about what the future might hold. After all, look what had come my way during a period of severe depression with its hopelessness for so many. What a wonderful life and challenging opportunities I had enjoyed! And emerging from a very commonplace academic record in high school, a straight-A record in college after my first quarter at Southwestern.

What had happened? Was it the change in venue? The motivation of independence in every decision I made and action I took? The people I met—faculty and student body I interacted with? Or, simply the underpinnings of my unique childhood exposure to a loving, understanding family that expected more of me than I was ready—or, perhaps even able at the time—to provide? Maybe a little bit of each. But as the years passed and maturity began to take hold, even as creativity began to wane, clearly the primary influence was the strong and wise people who touched my life along the way that made the lasting difference rather than the venues.

Southwestern, however, was unique in this regard. First, it offered a location for education concerned more with the search for understanding rather than just a storehouse for book knowledge. Second, its goals were structured more toward revealing and amplifying the strengths of its students and steering them in constructive growth, rather than for maximizing the number of A students throughout all their course work. Accordingly, the faculty was often selected less on the basis of scholarly papers published than their interests and abilities to help students frame intelligent unanswered questions about the disciplines studied. In my view and experience at Southwestern, this philosophy of education produced more scholars with worthiness as deep thinkers than doubting Thomases in probing the depths, frailties, and incompleteness of knowledge. That expresses why, to me, Southwestern was unique among small colleges in the United States, and is so even today.

I must make it clear that while SU changed the course, tempo, and purposefulness of my life, my growth and such achievement as may have been identified with my numerous other venues were again closely influenced by fortunate associations with a succession of people with a bent and talents similar to those who were so influential at Southwestern, a few whom I will list here and add a note or so about each describing what made them special to me.

Edward and Janice Onstot

I have always regarded Ed and Janice Onstot, who I will later discuss as role models, not only as the ideal married couple but also admirable people individually with multiple talents. And certainly NOT incidentally, I counted both among my dearest and most valued friends of all times. Both were graduates of SU, and he was a professor of psychology as well as the director of the band and orchestra who had brought me to Southwestern. From the very first, they both became not only mentors to me but steadfast friends and benefactors.

Ed Onstot provided me with a music scholarship to begin with and a steady job, first as dishwasher in the dining facility kitchen. Soon after, I was promoted to waiter. To these were added, three times a week, jobs in the grade school, and later in the high school, teaching group lessons in instrumental music. Finally, it was arranged for me to travel to Taylor High School, some 30 miles away, twice weekly to conduct beginning band section classes, three different sections each evening. Ed lent me his Buick sedan to travel for these classes as well as to attend local music competitions.

Ed had lovable and talented parents, his mother an acting mom to numerous students, three at a time, including A. C. Hart, Hal McComb, and myself, during our first year at SU. Ed's dad, Howard, president of Georgetown's principal bank, and an outstanding tuba player in the college band, was special to me for several reasons besides his contributions to the band, mainly because of his action the day I came to the bank to make a small deposit just before the bank closed. He called me over to his desk for a short chat before making my deposit. He kept me with small talk until I discovered the teller windows had closed, and then he said, "I'll take care of that (deposit) for you if you'd like." I thanked him and left as the bank's very last customer. It never again opened. But two days later, my deposit was returned in the mail with the terse note, "I thought you might need this sometime." Howard Onstot left Georgetown to become affiliated with an Austin bank soon after, and a few months later he was followed by Ed and Janice, an enormous loss to Southwestern and the community at large, but their leaving generated an incredible boost to my career, as we shall soon discover.

Louise Dicken

Having survived high school (and my first steady employment away from home in San Antonio, as well) without a steady girlfriend or many dates, I reached college with the reminder that with a college education and career of some kind ahead, plans for marriage should follow closely. I would not want to leave the impression that for me puberty was late in arriving, or that the customary sexual drives that it stokes were feeble or absent. Certainly NOT! But the need for exerting the discipline necessary to control them had been well explained by my dad, whose advice I respected and generally followed. But within my first few weeks at Southwestern, I met Louise Dicken, a talented violinist. Both Louise and I had been taken under the wings of Ed and Janice Onstot. Instantly aware of both our bonds in music as well as obvious personal attractions, we were continually invited into their home on the edge of the campus not only for dinner but to listen to Ed's library of fine music; to receive informal instruction on various techniques of music conducting; and, perhaps more important, comments (never didactic instruction) on things we needed to know if—more like WHEN—we intended to marry, lending us erudite books with both guidelines and admonitions.

King Vivian

The president of Southwestern was King Vivian. From a liberal academic background in the ministry of the Southern Methodist Church, Vivian came to SU at a critical time in its history when it struggled to keep its feet on the slippery slopes of the Great Depression. Enrollment was under 500; most students required some manner of support to remain, and the faculty's survival soon depended on salaries paid in script. Much of the school's operating overhead for services was defrayed by student work assignments, including management of the large farm adjacent to the campus, which supplied a major part of the food for the dining facilities. Such activity just to keep the school "alive and well" didn't just happen—it was doggedly inspired and driven by Vivian with cooperation of the equally driven faculty. Vivian was basically a liberal-minded scholar, but when pressed to do so, he became a very effective entrepreneur and hardheaded businessman. While highly respected by most of the student body, he was rarely observed visiting classes underway, or for sponsoring or participating in student affairs or activities. Students, in my recollection, tended to view him as less of a father figure than as an erudite scholar with remarkable skill as an orator. He displayed this twice weekly at the (mandatory) college assembly proceedings during his brief, but never trivial, remarks, usually ethically oriented but nearly always ending with a question to contemplate as you left for the next class.

One afternoon in late fall of 1931, I was surprised, and perhaps a little alarmed, to be summoned to Vivian's office. Students were rarely invited there unless something serious was involved. This time it was! He confided, "I've got a serious problem, and I need your help." With my mouth wide open, I listened as he quickly continued, "Ed Onstot has just informed us that he wants to leave Georgetown at the end of December to work with his dad in Austin. Of course he, as well as the rest of us, realize what a critical loss this will be to the faculty, coming at the time this does. It will take more than a year to fill this vacancy and we can't afford to shut down our instrumental music program that he headed with such distinction. But he tells me he is confident you are well equipped, talented, and experienced to guide this program effectively during our search for Ed's replacement. Are you willing to help us? We can't pay much, as you already know, but we'll defray all your basic living expenses and the cost of your continuing course work here." Of course it wasn't a question—it was an offer that couldn't be refused.

Ed may have stretched his representation of my abilities more than a bit, but I welcomed the mantle tossed to me and gave the opportunity my best shot. Whatever the assessment of my efforts by others, certainly without the opportunity afforded and the addition it supplied to my curriculum vitae, I would never have made it to Emory, and what followed from there. For all these (fortuitous?) events and favors, I'll be eternally grateful to both King Vivian and Ed Onstot. What a tremendous effect they each made on my life and career.

Vernon Guthrie

Guthrie, a professor of physics at SU, was a tall, lanky, soft- and pleasant-spoken person but authoritative in manner and appearance, a self-sufficient bachelor, and a proud owner and exhibitor of a Pierce-Arrow touring car that was always in immaculate condition. In a crowd, he was the first to be spotted but not the first to be sought. But in a classroom or laboratory what he had to say was brief, to the point, and you got it the first time (or were in need of sleep). Most of all, and without seeking it, he convinced me that my career would have to be in some branch of physics, not because of the 101 class of subject matter it contained, but because of the unanswered questions he dwelt on and the challenges they posed. His lectures always seemed to venture into the frontiers of science at that time—for example, radioactivity and even to the more simplistic aspects of relativity. In short, his drives, as were mine, centered on exploration. In retrospect, I may have been somewhat further motivated competitively. By happenstance, not design, my music associate and very first steady girlfriend ever, Louise Dicken, had signed up for the same physics class, and I soon found out that I had a real competitor in the understanding and application of physics, as reflected in biweekly test scores. Clearly she had me bested in physics. But, in my arrogant opinion, with less objective evidence to support it, the opposite was true in music. But this never became a bone of contention, as she pursued a career in music as a violinist and I in physics.

William Gray

An elderly professor of religions, Gray was a scholarly and, in my view, talented, liberal-minded, and inquisitive teacher. He was also a devoted advocate of intertwining the definitions of faith and understanding in studies of most other religions—a steadfast Christian in doctrine while curiously identifying and debating the world's great religions, each with their own seemingly conflicted meanings or intent in some scriptural passages. His search for truth, as the terminus of a rewarding journey from faith to understanding, I found not only a motivating but compelling road to travel. It was so similar to the explorations in science, where despite remarkable progress, in the end one finds you "never get it quite right" and you set out once more in search of fuller understanding.

Laura Kuykendall

Kuykendall was the dean of women, and throughout the university was regarded as SU's hostess with the mostest. She was well acquainted with most of the students and graciously presided over nearly all social events. She came from a liberal-minded social outlook but was stern when it came to social graces and manners. Kuykendall presided over a student body banquet monthly that encouraged party dresses for women and black tie attire for men. Those who were unable to meet these proprieties were urged to attend nevertheless, although they were always seated separately, a practice that earned her sharp criticism from the board and others. But she knew how far she could press the curve and refused to change her practices, and in the end very few students failed to acquire some kind of attire that met the dress code. This was not yet the age of protest, and in my view it paid dividends—socially, individually, and collectively. Louise Dicken and I were called regularly to provide violin-trumpet incidental music for social events, which earned us the appreciation and numerous special privileges from Kuykendall. To us, and many others, Kuykendall was an icon of social grace and behavior, an example worth emulating. We loved her, and cheers were abundant when the board renamed the women's dormitory Laura Kuykendall Hall after she passed away of a heart attack.

Dean Henry Meyer

A professor of music, skilled organist, and director of choral music, Meyer was my much-appreciated benefactor. He was responsible for my first regular employment after earning a master's degree, at the Crockett, Texas, high school. He was a tall, gaunt, soft-spoken, and kindly man, though not an outstanding leader in his profession, but the kind of faculty friend you felt comfortable with, always displaying an interest in you.

Randolph Tinsley

Dr. Tinsley, professor of geology, was an elderly but sprightly teacher, an insightful observer who taught his students not only to be careful observers but also to ask penetrating questions about "how did it get that way?" His field trips through the Hill County of Texas generated a great deal more curiosity than the inevitable appreciation of the glorious colorful fields of bluebonnets and Indian paintbrush during the Texas springtime.

William Wapple

"Uncle" Wapple, a professor of mathematics, was a wee gentle man with a high-pitched wee voice, certainly not a scholar, but a convincing teacher who must have set me off on the right track in my earliest exposure to calculus notation and equation building. At least my grades under him indicated we were on the same page. He was a memorable icon of education at Southwestern because of his ability to communicate with students, who tended to be fascinated not so much with the knowledge he tried to impart as with his voice and manner of exposition.

Burgin Dunn

Among students on the campus, Burgin was one of my best friends from my earliest days at Southwestern onward and, as it turned out, one of my most significant benefactors in paving the way for my continuing education and career advancement. He was a worthy member of the band, a social activist on campus, an interesting and jovial friend to all his associates, an ambitious physical scientist, and in my perception, on his way to a notable career. Burgin, son of a Methodist minister in east Texas, was one year ahead of me at Southwestern. After graduating from SU in 1931, he went on to Emory to pursue a master's degree in physics. During our communication exchanges, he learned I had no prospects of getting a job in Texas after my own graduation, and wrote and suggested, "Why not come here and finish your masters in physics with me? I can probably get you a scholarship to help out." After further exchanges about costs, I expressed my gratitude to what he hoped to do for me, but I still couldn't finance the balance needed for my academic and other fees. A week or so later, he called from Atlanta with the good news, "Come along ole buddy, I just got you still another scholarship...IN MUSIC." That did the trick, and it paved the way for my academic career toward a master's in physics.

I'll forever remember the Christmas holiday we spent together in 1934, when we hitchhiked rides from Emory to Nashville, Tennessee, to visit Burgin's elder sister, an editor in the production of the Methodist Church's widely distributed monthly for Sunday school lessons across southern Texas. After a delightful week there, we again hitchhiked to a plantation in southern Georgia, where the daughter of its manager was the girl Burgin was deeply in love with and expected to marry when he left Emory. But he didn't, because she rejected him at the last minute. How could she summarily reject a guy like that?

We shared a year at Emory together before we parted ways physically, but we remained in touch by mail, telephone, and during the annual meetings of our band members on campus for the remainder of his days. Burgin returned to Southwestern as an instructor in physics after graduating from Emory. There he fell fast in love with the daughter of Dr. Howard, an SU professor of English, who managed to scotch that relationship.

Burgin was best man at my wedding to my first wife, Mazie Houston, in San Antonio in late 1935. And many years later, my third wife, Joanne Gerould Starr Malkus Simpson, and I spent a number of happy reunions with Burgin, who ended his days at Texas College of Arts and Industries (now Texas A&M University–Kingsville.) A couple of academic misfortunes sapped Burgin's career ambitions but not his curiosity and capacity to inspire students. He was highly regarded as a classroom teacher, and he eventually was awarded emeritus status there and had both a ham radio center and a scholarship in physics named for him. Through it all, he was a very notable friend to so many of his associates, students, faculty, and others—especially myself, who had so much to be grateful from his enduring friendship.

These were the people at Southwestern who made a difference in my life and education as well as the career I sought, not through providing a warehouse of knowledge for my edification, but helping me to discover myself, my strengths and how to apply them; to reach out for goals and for means of their fulfillment, not by simple acceptance of knowledge, but to frame and pursue questions at their frail edges, where they were worthy of a closer look in their application. Southwestern was continually a venue of exciting exploration and investigation away from the hustle and bustle of city life and the generation of wealth. It was a place abounding with exciting and meaningful history, and with social and scientific curiosity, and in general a venue uniquely able to help you discover and better understand the world you would most love to be a part of, and to contribute to its betterment.

The End of My First Romance

During my first years at SU, Louise Dicken had been my constant and devoted companion, both in music and science as well as social activities. As strange as it later seemed, we never became intimate but looked forward to the time we were financially able to marry. But with no job in sight for either of us, it was not the time to throw caution to the winds and yield to our passions and need for togetherness—at least that was my view, though NOT Louise's. After Louise graduated a year ahead of me, she returned home to Duncan, Oklahoma. I hitchhiked from the campus to Duncan once or twice a month to visit her, but the stress proved too great. When she returned to the campus a little later, she returned the high school graduation ring I had given her a year earlier. I always felt it was a break made with shared, but not lingering, regrets. I learned later that Louise managed to gain acceptance at the Juilliard School of Music in New York while I launched my career in science at Emory.

During this time, I had met and become close campus friends at Southwestern with Nancy (Enid) Avriett, whose company and friendship I enjoyed during across-state tours with SU's glee club. Nancy was an SU graduate who preceded me in graduate studies at Emory, returned to SU as an assistant librarian, and acquainted me with what I could expect to encounter at Emory.

A Challenge

The dining hall at SU was surprisingly free of small talk or trivialities—which is probably how conversation one noontime turned to European culture, history, and politics. As always, the tables for six or eight were insofar as possible assigned to an equal number of boys and girls, primarily, we were told, to promote diversity of conversation and competitive points of view on a variety of issues. I cannot recall the issue that started it all, but a controversy arose between the girls and boys that was irresolvable with the information available at that table. The stalemate ended with one of the girls challenging the boys to spend a summer in Europe and to find the truth of the controversy. It sounded silly to most of the boys because we all knew none of us had the resources for such overseas travel. However, before dessert was served, one of the three boys at the table—Joe MacAuliffe, Ellis Wood, or myself—spoke up and remarked, "Well, sounds like a good idea, WHY NOT? We could form a small dance band, play our way over and back on one of the cruise liners for passage, and use third-class rail fare (cheap as dirt) to travel around France, Germany, and Italy." The laughs that that generated, considering the uncertainties and with less than three months to make such unlikely arrangements, seemed to end the discussion. However, at dinner that night, I suggested the three of us boys get together and reconsider whether the girls' taunt might have some merit. The attitude of the other two was if we could find two other band members who could scrape up $100 or more for the travel and were willing to risk hitchhiking to and from New York, it would be worth a try—nothing risked, nothing gained. The boys asked me, as (then) director of SU's Pirate Band, to explore the availability of "play for passage." To make a long story short, I did. It not only was possible but the Cunard–White Star Line agent was in Dallas and would like to audition the orchestra some time the following week! We had no such band or dance music available, much less time to rehearse for the audition. But I stuck my neck out, replying to the agent that one member of the five-piece dance band was out of town (indeed the truth), but that we could provide an audition the following Tuesday. It was a busy week, but with the help of a few friends in Austin, I recruited a pianist from the University of Texas who could sub as a drummer, another pianist from SU, and completed the five-part arrangement for several dance numbers for what we had decided to call The Texas Rangers. We adopted as our theme song the tune of the same name. The audition went well and after two dance numbers and the theme song, we were signed up for trip to Europe on the RMS *Franconia*, leaving June 30 from New York and returning late August from Cherbourg, France, on the RMS *Aquitania*. It was agreed we would make our way to New York separately and assemble at a YMCA there at least two nights before the scheduled sailing.

Ellis's parents, a Quaker family that had driven to Georgetown from their home in Avondale, Pennsylvania, to attend his graduation at SU, invited me to ride back with them to Avondale—that solved my transportation problems to New York. And the timing suited their plans to stop off in Chicago, Illinois, to take in the World's Fair, an unexpected treat. After spending several days with the Woods in Avondale and being introduced not only to the Quaker family's religious protocol at home and at church but also Avondale itself, then known as the Mushroom Capital of the World, they finally drove Ellis and me to New York City and to our accommodations at the YMCA.

In New York, I breathed a deep sigh of relief to find the remaining three band members already registered. The next morning, all five of us checked into the Cunard Line's New York office and received instructions for boarding the following morning. On advice of the Cunard agents, we decided not to change our dollars for British pounds before leaving the United States, but to wait till we arrived in London, United Kindom, because we were told that it was doubtless we would receive a more favorable exchange rate. This advice generated our first calamity! While we were en route to London, President Franklin D. Roosevelt (FDR) had taken the United States off the gold standard, so when we reached London, instead of the pound costing us $3.50, we had to pay $5.05. In a second blow to our plans, we learned that rail transportation costs had increased beyond our ability to pay, even in third-class accommodations.

The details of this journey alone could easily fill a fascinating volume, but they will be sketched here only briefly. It was an unlikely experience, instigated from a taunting dare by three female tablemates during a lunch at Southwestern, a dare the boys had lightheartedly accepted and somehow brought off. It was an adventure extraordinaire, fraught with uncertainties and obstacles that a less stubbornly determined group would have been tempted, if not forced, to abandon. It was the memorable experience of a lifetime. The following résumé of encounters and experiences along the way is probably worth including here, considering the handicaps, challenges, and rewards this journey served up.

Eastward crossing on the SS Franconia

Our accommodations were comfortable, in the tourist (or second) class area. However, we had the run of the ship during the voyage, with musical entertainment being confined primarily to the cabin class and for tea on the Cafe Veranda deck. Weather remained generally good, but the relatively short moderate seas made it more difficult for some passengers to acquire their sea legs, particularly on the Cafe Veranda dancing area, where only three of the five band members were able to stick out the tea dance at 4 p.m. the first full day out. Elven Edge took over the piano position,

Joe McAuliffe remained with clarinet and sax, and I attempted to fill in the bass and snare drums with one foot and one hand, and a trumpet in the other hand. The three of us fulfilled our obligation of the first teatime session, which had been quite a challenge, with the vibration from the three propellers coupled with the slow but steep roll of the 90-foot beam and round bottom of the *Franconia*. From the third day at sea onward, however, the band in full complement didn't miss a beat, although noon meals in the dining salon continued to present a challenge. When seated at a port-side table, we were looking out the starboard porthole 90 feet away. First, we viewed the cumulus clouds above, then with the next spoon of soup our view was of the churning sea below. It was disquieting at best!

Bicycling through Europe the summer of 1932

In London we were guests at the home of Elven Edge, our Texas University recruit for the band. In discussing our financial tragedy with the Edges, they suggested we might consider shopping at the Saturday flea market in town and buying bicycles for such travels as we could afford. We did and were delighted to acquire four handsome secondhand bicycles in good shape, none costing much more than $5.00 (U.S. dollars). My bicycle was only the equivalent of $3.50. Thus began our explorations of Great Britain, Belgium, Luxembourg, Germany, and France, a total of 1,275 miles on bicycles. We would resell the bikes in Cherbourg for less than a dollar of what we paid for them!

After acquiring our bicycles and spending a day sightseeing, we set a course northward, wondering how far we could get before returning to Dover, United Kingdom, and the English Channel crossing to Europe. We spent an extra day in Oxford, where we were surprised to obtain lodging for a shilling a night at a residential rental for Oxford students (who, at that time, were on summer vacation). The owner turned out to be a board member of one of the colleges, and he insisted on showing us around the campus. He then invited us to afternoon tea, during which we exchanged stories of life on our campuses at home and he of his life at Oxford.

The next day we proceeded toward Stratford-upon-Avon, encountering the first impediment to our travel plans several miles before arriving there. We approached a steep roadway known as Sunset Hill, where road signs had warned of a dangerous turn ahead. Leading the group, I stopped the others, identified the warning, and then trailed the others who negotiated the turn successfully. But when I started down, I discovered, too late, that one of the brakes on my bike had lost a brake pad. I didn't negotiate the turn but left the road, crashing through a rail fence, crumpled my bike, and tore a muscle in my right arm. What now? All gathered around, puzzled as to how to cope with the situation. I was in no condition to travel. But the second car driver that approached us stopped to see if he could help. He asked no questions but hustled me into the front seat of his car, my damaged bicycle into the back, and described where the others could reach the two of us at the hospital in Stratford and then departed with me. Joe and Ellis found me at the hospital described by my benefactor, and a note describing where he had left my crumpled bike but no indication of where he could be reached. Again, the cash shortage loomed threateningly as my companions puzzled over what to do next. Then, always the optimist, I suggested, "Just find the hospital front office and tell it like it is (regarding our finances). We'll telegraph for additional funds." The hospital replied that I would be there several days before I could leave. It was best to wait to find out how much the bill would be. Joe and Ellis returned from the bike shop with the same suggested delay. Three days later, together we sought the expected bad news about our indebtedness. In stunned disbelief we heard the clerk respond, "Your bills have been paid in full by the person who brought you here." At the bicycle shop, the story was the same! Unfortunately, it was not until two months later that I was able to express our collective appreciation to our Good Samaritan in a letter to the hospital, urgently requesting that our letter be forwarded to this Christian gentleman whose name or whereabouts we never discovered (at his request, we were told.)

Meanwhile, during the time required for my recuperation at Stratford, Joe, Ellis, and our pianist enjoyed the annual Shakespeare Festival (at student rates). When I heard this, I insisted on another night in Stratford-on-Avon to enjoy the performance of *Twelfth Night*.

With this near-tragic, but curious, adventure behind us, and with our summer days rapidly slipping away, we decided to abort our plan to visit Scotland and began backtracking, as rapidly as my remaining encumbrances permitted, to the White Cliffs of Dover and the Channel crossing to Europe. With some gratification but concern at the time lost, we crossed Belgium and Luxembourg into Germany without incident, hoping to reach our prime target of Heidelberg within a few days. But the impact of one Adolf Hitler, even in 1932, was beginning to be felt as far southwest as the Luxembourg border. To our surprise and chagrin, we were subjected to an unsavory taste of it shortly after departing Luxembourg. After crossing into Germany, the four of us were stopped every 8–10 miles for checkpoint examinations of our passports and luggage, and successively more extensive questioning—frighteningly so, both as regarding the tenor of interrogation and our concern with the loss of time. After all, if we failed to return to Cherbourg in time for the departure of the *Aquitania* for New York, then we faced the loss of transportation home with insufficient funds to schedule alternative means. After a long evening of debating our options, all of us agreed to turn back to ensure our scheduled connection to New York. We did and, a few days later, once again breathed a welcome sigh of relief to cross the border into France.

But even after a second shortening of our planned itinerary, and imposing increased frugality on daily living expenses, our pooled resources were diminishing much faster than envisaged, and we still had Paris and its more expensive venues ahead of us. So, we ate less and gave up the idea of seeking cheap shelter in humble accommodations along the way. Instead, we sought out wealthier communities to request permission to sleep in their freshly stocked hay barns overnight. It worked, after we promised not to smoke in the vicinity of the barns. This also ended up giving us a better appreciation of the country and its people socially. Frequently, the farm families would bring us blankets and other comforts and joined us before bedtime for discussions of compelling and dominant issues and interests in both our countries.

Near-tragic incidents

Once again, however, we were beset by an impediment we could have done without. Nightfall overtook us one evening a mile or two out of Château-Thierry, France, after a strenuous day of pedaling without availability of haylofts or barns but an attractive grove of trees with a small inviting pond, ideal for a brisk swim and an opportunity to care for some postponed laundry needs during a bright moonlit evening. Shortly after midnight, with most of us sound asleep, I was awakened when I heard an automobile on the narrow road above us cut its engine. Why the stop here? There was no park or facilities. No other vehicle except our bikes and they were not visible from the roadway. I dozed off in uncertain sleep a bit later. I have no idea how long I had dozed when I was awakened again by rustling of the bushes and multiple steps retreating toward the road, an engine starting, and the roar of a car moving quickly away. I woke the others and we all rushed to where we had left the bikes with our freshly washed windblown clothes draped around them. At first glance everything seemed in place till Joe exclaimed, "The rascals stole my camera and…oh oh!…my passport was inside it!"

With that we were all wide awake and searching our own possessions. Ellis commented in a suppressed voice, "I'm afraid they made away with them all."

"Now what will we do for passports?" I added. Suddenly, I queried the other three, "Where did you last leave your passports?"

Stony silence.

Then Joe replied, "I always hang mine around my bike seat, but . . ."

Ellis mused, "Last I remember, I gave it to you after our last checkpoint in Germany."

Finally, I remembered that after three consecutive checkpoints, we had agreed we would alternate, with one person carrying all the passports in his camera case. A few seconds later, I found both my camera and all the passports wrapped within my freshly washed shirt draped around the handlebars. All the cameras except mine were gone, a disappointing loss. But that was nothing compared to the trouble we would have been in if we had not recovered the passports, especially if the incident had occurred at a checkpoint in Germany!

Paris to Cherbourg

Despite the upsetting episode near Château-Thierry, we continued sleeping in barns or out of doors, all the way to Cherbourg without a further disturbing incident. The exception was during our three-day layover in Paris for sightseeing and attending a memorable performance of Richard Wagner's *Die Walküre* at the Paris Opera House, where we were surprised to get student standing-room-only accommodations for six francs each (at that time, a lowly 12 cents in U.S. currency). Even for that lengthy performance, we were glad to stand at that price. I am compelled to add that that initial exposure to Paris left me with a happier impression of Paris than I experienced on following visits there some years later. En route to Cherbourg, we were so low on funds that we were obliged to remain on a virtual fast, at times living off the land, digging up potatoes and plucking grapes from roadside vineyards. Our last day on the bicycles found us asking curious farming families for something on which to subsist; this did not produce the bread we had hoped for, but mainly fermented cider, which did not contribute to our travel safety. It did, however, muster exceptional concentration and determination to protect us from harm's way. You better believe we were exceedingly grateful to board the *Aquitania* when it arrived and to make our way to the dining room full of wonderful Cunard-style food. Even two stormy days at sea were a welcome lark after some of the 1,275 miles of bicycling.

The return voyage to New York on the *Aquitania* had few new events that captured our attention except for the two stormy days. However, we were weary and travel worn, anxious to be on solid ground on the way home. Doubtless, we missed many fascinating differences encountered on the return voyage. The storm seemed new because on this 900-plus-foot cruise ship, there was little of the rolling we experienced on the *Franconia*. The narrower beam and greater draft and length caused it to cut through waves, shuddering as it went. From our—I think—E deck portholes, we had a spectacular view of the trough 15 feet below us to the wave crest far above us, with that frightful shaking during the transit.

My return to Texas

Once in New York, we separated and lost no time setting out on our separate hitch-hiking quests en route home. After I hitched my first ride westward, I discovered I had exactly $2.50 to get me to Texas. Fortunately, the man who picked me up did so with the understanding that he wanted to drive straight through to Oklahoma City, alternating the driving and sleeping with me. I accepted, and all went well until I was dropped off on the outskirts of Oklahoma City early one morning, but not before both of us enjoyed a sumptuous breakfast together (which he paid for). But there my luck played out, although not for a lack of a ride. The new driver, jovial but much too red faced, offered me a beer as soon as we were underway. As politely as I could, I refused, noting as he reached for his own beer, the half-empty carton of beer cans in the backseat. The guy, if not drunk, was in no shape to drive. However, there was no excuse I could drum up to persuade him to stop until he reached the Texas border after driving over hill crests at 70–80 miles an hour on the wrong side of the road a number of times. Finally, he sought a pit stop. When he returned I was nowhere to be found.

After my fifth consecutive ride, I ended up in Temple, Texas, where I found a classmate at home willing to put me up for the night. He also boosted my remaining cash of 50 cents by a dollar bill. It was sufficient to get me back to Corpus, where Mom and Dad welcomed me with open arms. However, not until later, when I found time to pen a few brief notes on what had happened that summer, did it dawn on me how unique a summer it had been since the three of us had accepted that dare of the three girls at the luncheon table that spring day at SU.

If my return home was both welcoming and joyous, it was also brief. After little more than a week, once again I was on the outskirts of Corpus trying to thumb a ride to Atlanta and the Emory campus. Thanks to the entrepreneurship and determination of my remarkable friend, Burgin Dunn, I had obtained a music scholarship to attend this university. But I only had a few days to reach Georgia before classes started.

Emory University

I was not prepared for the delightful vista the Emory campus displayed at first sight. Nestled in an extensive grove of Georgia pine, all buildings (at the time) were constructed of Georgia marble in the same architectural style, including not only the academic buildings and classrooms but the dormitories as well. The uniformity in campus design and placement of structures around the quadrangle, as well as the architecture, and the accompanying Glenn Memorial Church nearby was a remarkably pleasing icon of higher education at its best. The cost of all this, we were promptly told, was entirely underwritten by the Coca-Cola Company. Not surprisingly, Emory soon became known as the Coca-Cola University (although not by request, or design, much less by demand of the benefactor). With the design and construction costs covered in the original grant, overhead costs of operation were low, and consequently tuition was lower than most universities of comparable stature. This also was apparently the answer to one of the early questions most students asked: Why was there such a high percentage of students from prominent families, not only from the conservative, money conscious but wealthy New England and the populous northeastern states?

Emory, from the very first, established its prominence and reputation in chemistry and medical science (along the southwestern side of the main quadrangle) and theology and law (along the northeastern side). The small physics and mathematics departments were clustered close by chemistry; the splendid library was on the western side, while the administration was on the eastern side of the quadrangle. Music occupied portions of the church and a few minor structures to the northwest.

At Emory I earned my keep (and paid my tuition) through scholarships—one in physics (my sponsor being Dr. Nelms, a Southwestern graduate and a professor of physics) and the other in music, working with Dr. Malcolm Dewey—and through singing in the internationally known glee club, which toured Europe every other summer under British sponsorship. I also served as assistant conductor of the small university symphony when a performance involved both the orchestra and glee club. Music had served me well at SU as a pleasant recreation and a source of essential cash at Georgetown, but Emory was a different story. It turned out to be drudgery, not recreation, and left precious little time to get on with my research for my master of science degree and thesis in piezoelectricity.

Emory at the time could not brag of its Nobel laureates (none at the time) or its outstanding social opportunities, for which some other Atlanta schools were known. But it was known for its doctorate-level careers. In physics its star professor was J. Harris Purks, a notable young graduate of Columbia University, New York, whose graduate research involved cutting-edge science in double-X-ray spectroscopy. I found in him someone able to turn me on to science and motivate me as no one I had ever met at that time could. It was he who interested me in the new science of piezoelectricity as a forerunner of sonar technology that I so wanted to pursue. My thesis provided the excitement that my musical involvement had failed to do. I decided to pursue doctorate-level research in the exciting new technology known as sonar.

One break from the drudgery was provided by the special performances at the Civic Music Hall, where glee club members had the option of ushering Thursday nights (in black tie attire), providing they promised to serve for every performance. This was a true privilege, because in those days the Metropolitan Opera spent a week in Atlanta each spring. In addition, there were many other performances of internationally famous soloists and ensembles. Despite the good fortune I had enjoyed as a selectively trained instrumentalist in Corpus and Southwestern, it was clear how far I had to go in music education to qualify as a true professional. I also appreciated how long the road to recognition, if at all, would be and how small the monetary benefits.

I couldn't help recalling my disappointing performance as a trumpet soloist in a state music contest at Temple, during which I lost by three points to a high school senior from Beaumont. I lost to a high school kid as an SU graduate and as director of Southwestern's band. True, his performance was flawless, and I had briefly broken a single high note near the end. No, that wasn't the career for me, although I wondered how many times the winner had made such an error during a performance. But my embarrassment and loss of personal confidence was fairly well assuaged sometime later when my close friend, Lonnie Newsom, a fine tenor saxophonist from Corpus, who had joined me at SU in 1931, told me his dad, who had learned of my dismay at losing to the Beaumont lad, pricked up his ears and said, "I'm not surprised. I know his family well; they tell me he has already been offered jobs in several big bands as soon as he graduates from high school. His name is Harry James!" That helped, but I had already had enough contact with professionals who had come up the hard way to realize a professional music career, as much as I enjoyed music, was wonderful so long as I had the choice of taking it or leaving it, but was not my cup of tea as a profession. I was willing to concede I was not good enough, or competitive enough, to take the hard knocks and uncertainties and to continue to enjoy it as I had in Atlanta.

Dedication of the Warm Springs Foundation

Among my memoirs of Emory, one unanticipated encounter, however, stands out above all others. One spring morning Dr. Dewey called my laboratory to ask me to have lunch with him. Something very interesting had occurred that he wanted to discuss with me. He said he had just learned President Roosevelt planned to preside at the dedication of the new Warm Springs Foundation Institute the following week, and Emory had been asked to supply the customary and special music for the occasion. Emory's president had contacted Dr. Dewey and insisted we pull out all the stops for the occasion. Malcolm said he had no idea of the protocol involved but asked me to check it out. He would have the glee club prepare three special performance numbers and our usual Negro spirituals, for which our club was well known. I should conduct the orchestra, providing any ceremonial music required, including the usual "Hail to the Chief" and processional music if desired. Somehow we both brought it off acceptably, if not in the Navy's accustomed steady-as-she-goes tradition, after several phone calls to the United States Marine Band conductor for guidance and suggestions. The climax of the evening (for the Emory group) was NOT the music, but what followed.

The dedication ceremony was carried out without a hitch, and as in nearly every Roosevelt appearance a feeling of electricity was in the air. After the ceremony was over, Roosevelt, again on the arm of his son James, retired to his quarters adjacent to the ballroom. However, as soon as the audience had left, and as suggested earlier by James, the two of them reappeared, walked across the ballroom, and graciously shook hands with each member of the orchestra and glee club and then asked Dewey and me to join him for a chat in his quarters, which of course we did, trying to suppress the exuberance we felt at this turn of events. Once we were seated, James excused himself, leaving the three of us alone for at least half an hour before James reappeared. Then, with all the joviality that we had anticipated from newsprint pictures, FDR launched into a flood of questions about the country and its future as we saw it. I was very flattered that he directed almost as many questions to me as to Malcolm. The conversation eased and we became more comfortable and informal—so much so that I proceeded to describe the fix I and my companions found ourselves in on our bicycle tour of Europe when we reached London and found ourselves too short of pounds because his taking the country off the gold standard lowered the currency exchange rate. This drew a big guffaw from the president.

Without a doubt this was one of the more memorable days of my life. And it confirmed for me once and for all times the greatness of this man and his ability to steer the country back from deep depression to prosperity, and restore the near loss of faith in our country and the survival of our form of democracy.

The baton

There was only one downside to this memorable occasion. When FDR returned to the ballroom to invite us to his quarters, he sat down momentarily on one of the orchestra's chairs. There was a startling "c-r-a-c-k!"—I knew immediately what had happened. My prized and historic orchestral baton had been ruined. In folding the music stands for departure, someone had removed the famous baton from the conductor's music stand to a nearby chair. How could I have left the baton unattended, even for a few moments?!! I was feeling sick all over, but I tried my best to make light of the incident; after all, it was only a conductor's replaceable baton. Replaceable for a couple of dollars, or was it?

I was given the baton as a special gift from the father of my SU classmate, saxophonist Lonnie Newsom. I was told that the Newsom family had acquired as a family heirloom a piece of walnut furniture (allegedly) belonging to and used by President Andrew Jackson. It had been stored away for many years because of problems in restoring it to a serviceable usage. Mr. Newsom, an amateur wood craftsman, decided to put pieces of this heirloom to use by carving various articles from the original walnut pieces, one of which was the baton presented to me at SU. I used it only on auspicious occasions at SU, and it was in use with an Emory performance group for the first time at the Warm Springs affair, where its historic excursion came to an end. FDR never learned of this incident.

Attempts to Continue Sonar Studies at the Doctorate Level

As I completed my master's thesis, I began planning to obtain a doctorate in physics. My ambitions in this regard, however, came to a crashing halt when I set out to obtain funds or scholarship support for this effort. The Massachusetts Institute of Technology (MIT), the University of Wisconsin–Madison, and the University of North Carolina at Chapel Hill were the principal leaders in sonar technology. My first letter of application was to MIT, which responded curtly that it was not the policy of the university to award doctorate-level scholarships to applicants whose previous academic training had been from universities in the South. Wisconsin had already awarded the available scholarships, and North Carolina placed me second in line for the only remaining scholarship. Then lowering my sights for employment, the only offer I received was from a college in Georgia to teach science, earning a salary of less than $800 annually. I ended up back in Texas with a high school teaching job that paid $1,275 a year. Quite a comedown from what I had expected with a master's degree in physics; but as I recall, a discouragement it was not. In no way was it the end of my world or the world I had sought. Or had yet to discover.

As a master's degree graduate of Emory

Mom and Dad, bless their hearts, made the long drive from Corpus to Atlanta to be with and honor me as I received my master of science degree in physics from

Emory University early in June of 1935. It was not an easy trip for them, especially Mom, and they found their pallid son still weak from food poisoning. After the graduation services, Dad ushered us all around Atlanta and the vicinity, displaying his considerable knowledge of Civil War history and of the roles of the principal participants involved. I was proud of Dad's extensive knowledge and familiarity with history but was embarrassed that it far exceeded my own, a trait evident in the trip home through Tennessee, visiting the Rainey branch of the family in Brownsville.

En route home from Atlanta, I had time to reflect on the last few years. The farther we got from Emory, the more preoccupied I became with what lay ahead. The main frustration was about marriage and the establishment of a family. This was a goal that had been instilled in me by Mom and Dad, not by preachment but by dinner table conversations about friends or acquaintances, whose successes or happiness in life somehow depended on an early and successful marriage with children on which they could be proud. If it was adult conversation, then it was not lost on the curious ears of the children as well as the teenagers present. It was indeed a different age; the discussions or queries about sex and its role in family life, or beyond, was unthinkable table conversation, and rarely elsewhere (more's the pity, in my current view).

I frequently inquired of myself if my judgment had been terribly flawed when Louise and I broke up in 1932. I was to find the answer as soon as we got home. As customary returning from a trip, I called the Stant Simpson residence in Corpus to find what was up. Gladys, who had just graduated from Southwestern, answered and before I could launch into my account of the ceremony at Emory, she excitedly broke in, "GUESS WHAT! Louise is here. Why don't you run over?" My answer was obvious, "I'm on my way!" When I arrived a few minutes later, Louise's first comment from her perch alone on the front porch swing was more of a statement than a question, one I was unprepared for: "And how is Nancy Avriett?"…WOW! It was a warm night, but the atmosphere on the porch swing was ice cold. Perhaps I deserved that—I'm not sure, but I certainly wasn't ready for it. But ready or not, and frankly disappointed as I was, strangely I survived the jolt without scars or bitterness and turned the corner with no lasting regrets.

The rebuff I had gotten from Louise upon returning home from Emory, while disappointing, was short of discouragement. I was left with a feeling of emptiness, still with no option in sight to meet the challenge—or was it the need?—to seek an effective marriage. Whatever the chemistry or other factors influencing my physical drives and decisions, marriage, at that point in time, was accorded the higher priority, although it was certainly not far behind that of a career. Just how flawed my judgment was at this juncture is perhaps best illustrated by what happened next. I proceeded to get in touch with Nancy Avriett and less than casually broached the possibility of rekindling the flickering flames we enjoyed at Southwestern. However, as I should have easily suspected, her answer was NO! However, not to worry, I reasoned, and began planning to meet the challenges of life as a high school teacher in the small but venerable town of Crockett several months later.

Chapter 3
My Early Career

Crockett, Fort Stockton, and Corpus Christi

I accepted a position as band director and physics teacher at Crockett, a venerable and charming "Town of 10,000 People and 20,000 Pecan Trees," as it then proclaimed at its city limits. The next five years carried me from Crockett across the state to Fort Stockton, and finally back to Corpus Christi, mainly as the result of successes in music, but with no opportunity in sight to return to science. However, in 1940 that opportunity did finally show up when I was offered employment with the Weather Bureau at Brownsville, Texas. Here, music once again became a rewarding avocation and science a preferred career involvement.

Mazie Houston

Back in Corpus Christi in early June 1935, my Uncle Clyde Rainey, a foreman in the San Antonio Printing Company, recently divorced from Aunt Edna, called Mom to say that he planned to spend several days in Corpus on vacation and wondered if he could overnight with us at 1223 Second Street. He added casually that he would also be transporting two young ladies from Travis Park Methodist Church, Mazie Houston and Mary Shultz, who were also planning to spend a few days in Corpus. They had reserved accommodations in a tourist cabin on North Beach, but he wanted to introduce both of them to our family. As I now recall it, Mom immediately agreed and, as customary, sprang into action, making detailed plans for all, volunteering my services to introduce the girls to Corpus Christi while she prepared for a picnic party at nearby Aransas Pass beach. And, as usual, her plans kept us all busily entertained through the weekend.

Actually, I had not been overjoyed to have my weekend reprogrammed, as I had already scheduled activities with local friends. But when Uncle Clyde arrived and introduced his two passengers, my attention remained fixed on only one of them. Very pleased with what I saw, I quickly forgave Mom for her presumptions on the short time I had at home. I found Mazie an exuberant and attractive personality, a good conversationalist, and a person I felt comfortable with almost immediately. It turned out that Mary Shultz had others she planned to visit in Corpus and so she opted out of the next day's excursion to the beach. Once the rest of us arrived at Aransas Pass, Mazie and I found it convenient to walk the beach alone, becoming reasonably acquainted with each other. Then we enjoyed a swim together before rejoining the others for the beach picnic.

Thus began a whirlwind flirtation that led to a second visitation with Mazie in San Antonio the next weekend. Before the weekend was over, we were engaged with plans for a Christmas wedding after I had settled into my new job in Crockett. Whew!! In later reflections, I couldn't believe so much had been telescoped into so short a time, nor did I realize into what dangerous waters such haste could lead us. It didn't occur to me until much later that while at Southwestern, if I had had a permanent job confirmed, Louise and I would probably have been married, and I never would have gone to Emory, nor have ended up with a career in science. But it isn't my disposition to ponder such "What if?" questions at potential branches in the road of life. Rather, the purpose of my saga is to tell what DID happen that made a difference, warts and all.

To continue, Mazie and I met next on Thanksgiving weekend 1935 to complete wedding plans and to send out invitations to family members and special people, including my friends from Southwestern still on the campus and a few others with whom I had maintained correspondence, especially Burgin Dunn, my closest friend on campus; Joe McAuliffe; Ellis Wood; A. C. Hart; and Hal McComb. But not to Louise or Nancy, whose whereabouts were then unknown, and in truth I did not exert much of an effort to reestablish contacts with them. Why? I'm not sure. But then it was just too soon and too many questions remained.

Catching up with Louise

In any event, with our plans in place and Mazie having left for her plush job at San Antonio's NBC outlet, I was waiting for a bus to carry me to the post office to stamp and post our box of wedding invitations. A big Buick sedan pulled up to the curb and a voice from the car called out to me, "Hi, ole friend. Can I give you a lift?" Louise, expensively attired and jovial, seemed very glad to see me. What a jolt! Of course, I joined her without hesitation but no doubt with confused emotions.

Louise hastened to explain her plight. It seemed she had married a Texas rancher whose deceased first wife had left him with two children. Then he promptly passed away, but not before he had managed to change his will, leaving Louise as a non-beneficiary without immediate access to any resources. So, she left his children with other family members who were beneficiaries. Obviously, in the few minutes it took her to relate this sad episode I was much too flabbergasted to break in to learn further details. But after a pause to catch her breath she queried, "Tell me what's happening with you, and where are we headed?" After a considerable pause (probably with my mouth hanging open) I replied, "It's a long, yet short story. Can you guess what's in this box of cards?!!" About 15 somber minutes later, Louise dropped me off at the post office. It was a silent but emotional departure with tears in both our eyes, but without overt evidences of the sadness I believed we shared at that moment of frustration and uncertainty.

It was more than a decade later when next I saw Louise during a brief encounter at Southwestern. She was a much-changed woman, so much so I turned to A. C. Hart, one of my oldest classmates, and whispered, "Who is that young lady who just joined us?" He stared at me in disbelief and whispered back, "Why, that's Louise!" She was indeed buoyant, jovial, and in high spirits. In chatting with Burgin, she spoke of her experiences at Juilliard Conservatory in New York and mentioned that her new husband was an executive with an oil company based in Amarillo, Texas. She was now Louise Conway.

We greeted each other casually and chatted pleasantly, with no outward evidence of tension, trying, futilely, to appear comfortable with each other. Then she left to visit with other campus visitors. I felt we both parted with somewhat lighter steps.

The last time I saw Louise, and the only time I met her husband, was in 1964, when on a visitation trip to research forecaster offices. It was convenient to overnight in Amarillo while en route from Denver, Colorado, to Fort Worth, Texas. I called Louise at home and invited her and her husband to have dinner with me at my hotel. She said she was just preparing supper for the two of them and asked me to join them at their home, which I did. Louise, who was showing her age prominently, did most of the talking; Conway himself was courteous but notable mainly by his silence, and Southwestern did not enter the conversation. I excused myself early, returned to the hotel, called Joanne at home and described the strangeness of the evening to her, and then turned in, sleeping peacefully through the night.

While I never saw or contacted Louise again, I was surprised when Joanne told me one evening of a call she had received from Louise during the day, in which Louise commented, at least briefly, on the Southwestern days, hoping that Joanne and I had had a happy life together, and adding she was certain she would have had a much happier life if it had been with me (or words to that effect). Some months later, I learned from a Southwestern alumni publication Louise had passed away. Startlingly, it called to mind that within days of the published date of her passing, my dreams had been filled with memories of our exploits together at Southwestern, dreams I could not recall ever having within decades of memory, or having discussed with anyone while awake. But it surely got my attention.

Music Director at Corpus Christi High School

By 1938, I had come home, hired by Corpus Christi High School as music director. Both my sister, Mayme, and my brother, Clyde Roy, were eager participants in band and orchestra at this time. Mayme was a first chair flutist and Clyde Roy a percussionist and drum major in the marching band. They were both a joy to work with and I had every reason to be proud of both, but I struggled not to show it too openly in front of other students to avoid the perception of favoritism.

But our association in this venue was short, as my career turned a corner when I entered government service in meteorology less than two years later. Their lives followed quite different courses from mine with few opportunities for the family and educational togetherness that we enjoyed from 1938 to 1940. Nevertheless, each of their careers could fill a fascinating volume of fruitful achievement and in their own separate ways ended up making an important difference in the lives of many with whom they interacted. I remain very proud of a bit of research Mayme and several friends carried out, which showed that of the band and orchestra members I worked with in Corpus, more than 30 went on to successful professional careers in music and a number of which ended up in distinguished university departments of music. Mayme herself remained in public school music until her untimely death at age 57 (of a cerebral hemorrhage), a mother of four children: Susan Cates, Tommy, David, and Jim Boyd.

Clyde Roy, who enlisted in the Navy less than six months before Pearl Harbor, served in the Pacific Theater throughout World War II, and thereafter at Corpus Christi; Pensacola, Florida; and Norfolk, Virginia, until medically discharged with polio while I was still in Hawaii. After attending divinity school at Tennessee Temple College, he did Baptist missionary work in the Caribbean area, later transferring to Australia, Fiji, and numerous other venues. He served temporary assignments in overseas pulpits after his retirement, and is presently assisting in a ministry in Las Vegas, Nevada. With his first wife, Margaret Ann (incidentally, an oboist in my Corpus Christi orchestra), they had four children, and with her passing he had four additional children with his second wife, Joyce. And I thought I had served in a lot of different venues!!!

Another of my music students from Corpus Christi was a lanky, young trombonist by the name of Gilbert Clark. Little was I to know at the time how our paths would cross again some 20 years later.

The Simpson family grew a bit during these years. A few hours after Christmas 1938, while my high school band from Corpus Christi was performing its half-time maneuvers at the state championship football game in the Sugar Bowl at Dallas, Mazie gave birth to our first daughter, Peggy.

Brownsville

In early 1940, I received a letter offering me a choice of jobs in the Weather Bureau as weather observer at $1440 per annum, little more than a third of my income as director of instrumental music at the 12 public schools in the Corpus Christi district. After five eventful and curiously rewarding years as music director in Texas public schools, not by choice but necessity while in search of a viable opportunity to launch a career in science, I finally found some sort of fruition and accepted the position of weather observer and aviation weather briefer at the airport Weather Bureau office in Brownsville. It was a reckless gamble, made in desperation to establish a career in science. After all, I had a master's degree in physics and math and, in my view, had wasted five years of employment in music while my education in science was suffering from atrophy.

While it was rewarding to be back on course with a career in science to look forward to, almost immediately there was ample room to question the wisdom of the radical change that a career in science entailed. As a teacher in Crockett, my initial annual income had been $1275. But during the five years in public schools of Texas, that had grown to approximately $4400, including that from private music lessons and bonuses from the sale of musical instruments and equipment. But when I chose to become a classified employee of the federal government, my income dropped once more to a meager $1440 annually with no other source to bolster living expenses. This became the more unsettling when I discovered that the meteorologist in charge, after 25 years of service, had to content himself (and he was) with only $2000 annually. The bitterest blow, however, was the discovery that the Pan American meteorologists next door, with no formal education in meteorology and only a few years technical experience, earned $2200. It just wasn't fair! And Brownsville, in its isolation from the bureau's regional office in Fort Worth, not to mention the central office in Washington, D.C., which doled out promotions, offered little prospects of upward mobility. This was, truly, a grim outlook. All of which fed my resentment of the overoptimistic advice given me by J. P. McAuliffe, head of the bureau's Corpus Christi office, and father of Joe Jr., my close friend and roommate at Southwestern. In my concern about leaving a $4400 job for a $1440 one, he had not hesitated to assure me that with my educational background, the bureau would soon send me to Chicago for graduate training in meteorology and then there would be no stopping my advancement. Years later, reflecting on my adventures in meteorology with Joe Jr., I realized how wrong that advice had been, yet how very much on target it had turned out to be (though NOT in accord with the scenario J. P. had in mind).

It was a humble and instructive start in meteorology but one that, careerwise, led nowhere. Soon I was once again in search of a more attractive venue for advancing my vocation in science. Strangely, the opportunity arrived during an unexpected visit of Mr. John Riley, an elderly Weather Bureau official from the Washington, D.C., headquarters, to resolve a personnel quarrel at the city office in Brownsville. Incredibly, two of the three employees at the city office had preferred charges against their boss, an equally elderly sedate gentleman of retirement age, for dating a young Mexican woman! Riley ended up, however, spending more time with me at the airport than in resolving the quarrel downtown.

The outcome of my plea to Riley for help was an offer that in exchange for my acceptance of an interim assignment of several months at Swan Island (in the northwestern Caribbean Sea), the Weather Bureau would transfer me to the New Orleans Forecast Office in Louisiana as an apprentice forecaster with a grade promotion. It was an offer I couldn't refuse, my family willing (and it was). Coming as it did in early spring of 1941, of course the offer came without any foresight of what December 7 would bring, and the impact the war years would have on my newfound career in science.

Swan Island

A massive mountain rising singularly from the deep waters of the Bartlett Trench in the northwest Caribbean Sea, its summits several tens of feet above mean sea level (MSL), forms the two tiny islands making up the Swan Islands. One island (Little Swan) is uninhabited and virtually barren save for scrub brush and a few scrawny goats, and less than a half a mile in diameter. The other island (Great Swan), separated from the other by a shallow channel of wading depth, is a startling contrast. Approximately 1½ miles by approximately ⅝ a mile in size, Great Swan, at the time of our arrival, was covered with lush green foliage of hardwood—mostly manchineel —trees interspersed with tall coconut palms. In addition, the five residents were able to bask in the knowledge that Swan Island (as Great Swan was often called) with its lush vegetation, exists without annoyance or hazard from mosquitoes, snakes, or hazardous insects commonly found in most tropical habitats, except for the dragonlike and frighteningly large (generally 5–7 feet long at maturity), but totally vegetarian iguanas, a menace solely to the production of coconuts.

The weather station, a slightly elevated wooden structure, was situated near the island center about 150 yards from the living and dining quarters. The sole access to the island was by United Fruit Company ships that delivered six weeks' supply of food and sustainable, plus operational, supplies. These included 110-pound canisters of helium for sounding balloons. There was no powered facility for unloading supplies, only a tiny 30-foot sailboat (without auxiliary power), with which supplies were transferred from the circling supply vessel to a small pier, a task requiring more than a half day to complete.

Great Swan boasted a total population of five: one cook and four weather observers, who together were responsible for weather observations 24 hours a day and for communicating the observations by continuous-wave (CW) radio to the Navy in Panama for relay to Washington, D.C. In addition, they were responsible for maintaining the power and sewage plants; the water storage facilities; and the servicing and repair of the Bendix cycloray recorder, which tabulated the pressure, temperature, and humidity data from the weather sounding balloons. Finally, as time permitted, they maintained climatic records. As I view it today, these were daunting responsibilities, timewise and skillwise, by comparison with what was required at a first-order weather station on the mainland in those days (such as Brownsville), even without considering the requirement to demonstrate an ability to transmit CW radio messages at 20 words per minute to qualify for service at Swan Island.

But Swan, in its isolation, offered many challenges in appreciating and understanding the unique natural ecology of this magnificent island and the tropical environment that sustained it. Nevertheless, with the isolation, a much greater handicap for most observers was finding suitable involvements to fill the spare time. This proved to be less of a challenge for me, however, because of the six technical volumes on meteorology I had brought with me, and the urgency I felt to understand enough about my chosen profession to equip me to do well in my promised job as apprentice forecaster at New Orleans.

I was immersed in this secluded life when in the late afternoon of December 7, 1941, I had my CW radio transmissions interrupted by the Navy in Panama with the news that the Japanese had disastrously attacked Pearl Harbor, on Oahu, Hawaii, and that we were at war.

Francis W. Reichelderfer

My first encounter with Dr. Francis Reichelderfer was during my temporary assignment at Swan Island, where I had agreed to a six-month assignment in exchange for a guaranteed promotion and reassignment to New Orleans as a forecaster in training. When my six-month tour was up, John Riley, the man who had made the promise, had been transferred to Kansas City, Missouri, as regional director. Joe Lloyd, who had replaced Riley, didn't seem to know anything about the promise. The answer from Lloyd about my request for reassignment was, "Sorry, you'll have to remain at Swan Island indefinitely." Because of the war, it was his view that the bureau could use me best at Swan and, besides, they didn't have a place for me on the U.S. mainland. My radio response repeated the terms of my temporary assignment, my qualifications (academic), and that I had no intention of remaining at Swan beyond what had been promised me, not in writing but with a shake of the hand with a central office official, which any Texan would expect to be honored. His curt reply was, "See preceding message." My answer, dispatched less than an hour later, was both as curt and unyielding (to the effect), "Please refer to previous exchanges of correspondence. Herewith is my resignation from the U.S. Weather Bureau effective one week from today. On that date, I intend to board a schooner scheduled to call at Belize from which I'll find my own way back to Texas. [I refused to be further employed by an agency that could not be trusted to keep its word and solemn promises to its employees without the courtesy of explanation or discussion of alternatives.] Of course, I'll expect to support the war effort but not at the bidding of this bureau." Within the hour, I received a reply by radiogram signed "Reichelderfer, Chief of Bureau" without reference to any previous exchanges, confirming my transfer to the New Orleans Forecast Office with a double promotion, to grade P-1 from SP-4, and confirming my passage to New Orleans on the United Fruit Company transport SS *San Jose*. Needless to say, my response was in two messages: the first, an appreciation of the personnel action, the second, withdrawing my radiogram of resignation from the bureau, without reference to the earlier correspondence or the reasons.

Reckless or not, my message turned out to be a fortunate investment, not just for my future career, but because of an early encounter with a man who made a remarkable difference in the outcome of my life as a mentor and friend. He gambled on my ability to ultimately make a difference in the administration of scientific research and the prediction of severe storms, especially of hurricanes. Francis Reichelderfer was a man of extraordinary intelligence and leadership. In his 15 years of interactions with Congress and its leaders, he earned their respect, trust, and a reputation for impeccable integrity as a distinguished naval officer with graduate-level training in the Bergen School of modern meteorology. In his naval service, he was recognized for his expertise in weather prediction for lighter-than-air craft operation and for understanding of aviation meteorology. When appointed by President Roosevelt as chief of the Weather Bureau, he was charged with modernizing the civilian weather service, for which he quickly became recognized in the United States and abroad for effectively achieving. His reputation internationally led to his appointment as first president of the new International Meteorological Organization (IMO) and forerunner of the weather service arm of the United Nations, the World Meteorological Organization (WMO).

The story of my month-long journey from Swan Island to my new assignment at the regional forecast center on New Orleans's Lake Pontchartrain is a book-length saga but is only briefly summarized here. It was neither along a direct route nor uneventful. As directed by the War Department, the SS *San Jose*, a United Fruit Company banana transport vessel, picked me up at Swan and dropped me off in New Orleans and was en route 27 days. The ship turned out to be a transport mainly for war equipment, calling first in the Panama Canal Zone, then at Curaçao, and finally returning to Panama to load on the originally intended cargo of bananas. During these 27 days at sea, the radio operator aboard received 25 SOS or SSS distress messages from vessels in the western Caribbean that had been struck by enemy torpedoes or were being chased by enemy submarines. As we approached the Yucatán Channel to enter the Gulf of Mexico, the War Department reported submarine activity in the vicinity and routed us back toward Roatán, Honduras, with a day's delay before proceeding safely to the fog-enshrouded mouth of the Mississippi River and another tense 12-hour delay before proceeding safely to New Orleans. Several weeks later we were saddened to learn that the old *San Jose* had been sunk by enemy torpedoes while trying desperately to reach safety in the port of Belize. In retrospect, it is startling to realize that so much had happened in so short a time, in less than a year.

New Orleans

When my family arrived from Brownsville, New Orleans was abuzz with war-era industrial production, the largest of which was the Higgins Shipbuilding Company's incredible rate of constructing Liberty ships. In desperation, with no residential rentals available (except by under-the-table negotiations), we were obliged to remain in a tourist cabin near the shipyard for a month. With the help of friends, we stumbled on a humble row house for rent on Franklin Avenue. We had barely settled when, three months later, the landlady gave us one month's notice to leave, not because we were unsatisfactory renters, but because she had relatives coming to town who needed a place to live. Furious, I sought a lawyer who promised to invalidate the evacuation notice. But after 10 days he threw in the towel, advising it would take longer than the notice we had been given to make it through the courts. Again, with outraged help from the small Methodist church we had joined, we were able to move into a nice six-room house but with a catch-22. The Southern Pacific mainline tracks passed about 200 feet from the front of the house. There we remained until we left for the Weather Bureau scholarship program at the University of Chicago, just short of a year later.

New Orleans during that period was not the inviting place to visit, much less establish a residence, for which most Mardi Gras visitors like to remember it. Despite the problems we had in finding a rental home, the two things I remember and resented the most were 1) at Peggy's New Orleans school, there weren't enough desks to go around so she sat on the floor; and 2) the sailor language she brought home from school practically drove us nuts!

New Orleans had one of the most comprehensive and important weather forecast responsibilities in the nation, with separate specialized units for aviation weather, district and urban weather, river and flood forecasting, fruit frost and agricultural weather, and hurricane forecast and warnings—all conducted from the city office in the post office building until the late 1930s. Then all the responsibilities but river forecasting and the daily publishing of weather maps were transferred to the fine new airport terminal building on the shores of Lake Pontchartrain. The commute to work from our residence took more than an hour, involving transfers from bus to streetcar to another bus, and a half-mile hazardous jog through a crime-ridden neighborhood with risks about the equivalent of a 2008 nighttime stroll alone through northeast Washington, D.C., at midnight. But once at the airport, the environment and facilities were delightfully accommodating.

In less than two months of forecaster shift work at New Orleans, I was introduced to a new Weather Bureau employee entering (as I had in 1940) as an SP-3 observer with the same annual salary of $1440. She accepted this pittance out of necessity, since the University of Mississippi was paying her less than $1000 as an assistant professor of physics. Her name was Mary Hodge, a quiet but erudite young scientist, who I was astonished to learn had earned a doctorate in physics from the University of North Carolina at Chapel Hill, and was even more astonished to find out that she had been able to do so with the help of the scholarship I had applied for in 1935 after completing my master's degree in piezoelectricity from Emory. I had come in second among applicants and, with no other options in sight, I accepted the job as high school band director in Crockett. In August 1935, I received a message from the University of North Carolina stating that because of some default, the winner of the scholarship was unable to attend; therefore, the scholarship was mine, and when could I arrive for the fall session? Sadly, I had to respond it was too late to accept the scholarship because I was otherwise committed. Mary Hodge had been third in line, had earned the doctorate, but ended up having to seek government employment which she was much too overqualified for just to earn an acceptable standard of living with little likelihood of advancement. It just wasn't fair. And but for good fortune and a little help along the way, there go I.

The next day I threw caution to the wind and telephoned Dr. Reichelderfer and identified myself. He remembered me, listened, asked a few questions, and seemed from the tone of his voice to be interested. He asked me to write him, summarizing the incredible coincidences that I had described and then promptly hung up. I was left uncertain about the outcome but was convinced I had done my best to make a difference in someone's life—and as it turned out, I had. Less than a month later, a letter signed by Reichelderfer reached Mary, offering her a transfer to the Instrument Division in Washington, D.C., with a promotion to a professional grade. In her quiet and always understated way, she contributed significantly to the progress of instrument development, primarily in improvements in rawindsonde technology, before transferring to the University of Chicago as protégé and colleague of Professor Oliver Wolfe and his research programs in upper-troposphere disturbances of the midlatitudes.

This small episode is significant, mainly because it was so typical of Reichelderfer and his efforts not only to know the new staff coming into the bureau, their strengths, and motivations but also to have them realize their big boss was interested in them individually. And his staff was instructed that any employee visiting the central office for the first time not be allowed to depart before spending at least a few minutes with the Bureau chief.

University of Chicago (1943–44)

Less than two years after entering the Weather Bureau, I found myself at the University of Chicago's Institute of Meteorology, beginning a yearlong exposure to graduate courses in meteorology conducted by some of the world's most accomplished meteorologists. It was headed by the great Carl Gustav Rossby, whose influence was preeminent in introducing modern meteorology and its applications to the United States, and its practice to the military during World War II. I thoroughly enjoyed my time at the University of Chicago with Rossby and his star-studded staff.

There were 11 other scholarship students from the Weather Bureau, and hundreds of Air Force cadets and Navy officers, who participated in accelerated weather training programs at the University of Chicago. MIT; the University of California, Los Angeles (UCLA); California State University (Cal State); and other colleges offered similar courses, all mandated by the necessity of war and the need for forecasters around the globe. At Chicago, the academic pace of training in theoretical meteorology and weather prediction based on observations from a single station was awesome, challenging, and at times exhausting, leaving little time for appreciation and enjoyment of what Chicago itself had to offer

This was especially true for those of us whose careers depended on academic records. Despite the obvious care in selecting the 12 Weather Bureau students, as I recall, only two-thirds of the students came away from the Chicago exposure with careers of distinctions. Each Weather Bureau student had been advised well in advance he would probably not be returned to the station from where he had been selected for training. So, it was a tense moment when new assignments were read over the telephone from Washington, D.C., to the lunchtime gathering in a Chicago classroom. The outcome was placed in greater suspense by a group decision that each student would record on the blackboard, in advance, both his preference for and his prediction of his new assignment. Nearly everyone felt he had a good handle on where he would end up based on his interview the week before with Jerome Namias, who had been sent from Washington, D.C., to review the academic records and interview each student. As it turned out, of the 12 hoped-for assignments placed on the blackboard by the students, not one was on target! However, informally two students had indicated that they would probably seek employment in another agency if assigned to a particular office at which he wouldn't be caught dead. Roy Fox felt that strongly about Anchorage, Alaska, and Bill Burnet the same about New York. Their worst fears were realized, but after a year both were reported as delighted with their assignments. And both moved ahead to outstanding careers, Roy becoming a regional director and Bill a deputy Weather Bureau director. However, I cheated on my blackboard choices, hoping for Miami but recording Washington, D.C., as my preference. I ended up with Grady Norton in Miami. I never learned how many others "cheated" likewise or whether their outcomes were similar to mine.

In Chicago the family lived just a few blocks from the campus in a comfortable Kimbark Avenue apartment. When I had just completed the first quarter's strenuous courses, except for final exams the following day, Mazie, well into her ninth month of pregnancy, agreed she felt well enough to venture into the city to enjoy a performance of *Oklahoma*. Most of my Weather Bureau classmates had already seen it and raved about it, so we decided to catch it before Mazie gave birth. We awoke at daybreak the next morning, Mazie frightened with labor pains, only to find the ground was covered with 10 inches of snow. We had no car and only two of our Weather Bureau friends, George Kalstrom and Bill Burnet, did. When telephoned, neither felt he would be able to reach us without tire chains, which neither had. Finally, I decided to call the fire department for help, but we were concerned how long it might take to reach us. Then Bill called back, saying he had been out and decided he might be able, with luck, to reach us: give him 10 minutes. We did and so did he, reaching the hospital with skill and good luck in ample time. After a preliminary examination of Mazie, the attending physician advised it would be at least six hours before the birth. So Bill and I decided to rush to the classroom and take our exams and then return. But when we returned about two hours later, my second daughter Lynn was making her presence known to all in hearing distance. We found Mazie in good spirits, happy that we had returned without incident, and was pleased with herself. With all this exhilaration of the birth and the exams, I cannot recall a single hour of exhaustion.

At that happy moment, I couldn't help but wonder what kind of guardian angel had been hovering over us, considering the episode two months earlier, when the same hospital had advised us that Mazie had an advanced case of tuberculosis that would jeopardize the life of the baby, a misdiagnosis that was due to the hospital's misfiling of two X-rays. Lynn has demonstrated in her life that she is made of sterner stuff!

Victor Starr

There was one professor at the Institute of Meteorology, however, who still remains in my memory as the very best and clearest classroom lecturer and motivator of his students I have ever encountered—a small, soft-spoken, but effective, professor named Victor Starr. A protégé and close associate of Carl Gustav Rossby, who brought him to this country from Europe, Starr may have been quiet but he was a brilliant meteorologist. Starr tutored and prepared military cadets to apply the latest methods of "modern meteorology" to weather forecasting during World War II. Early in his career, Victor authored one of the first hardcover books on synoptic meteorology and weather forecasting published in the United States, exclusively based on Bergen School concepts of air mass and frontal analysis. When I was placed on temporary assignment to Swan Island, a copy of Victor's book and a new one by Jerome Namias, on airmass analysis, were the two textbooks I feasted on to qualify me as a forecaster. As it turned out, they were good choices. However, neither was in use or available in New Orleans when I was transferred there, but I continued to apply their new concepts experimentally there. When I was sent to Chicago, these proved to be a boon to my understanding of the class work there. After the war, Starr transferred to MIT as a professor in synoptic meteorology, where he taught for the remainder of his career.

Joanne Gerould Starr

My first acquaintance with Victor's wife, Joanne, was at the University of Chicago in late March 1943 at a seminar I had been asked to give. Both Joanne and Victor caught my attention because of their obvious interest in what I was trying to communicate about my first adventures in tropical meteorology while at Swan Island. While I frequently encountered Victor during the remainder of my year at Chicago, my next encounter with Joanne was considerably later, in New York at an annual meeting of the American Meteorological Society (AMS), where she was giving an interesting research report on cumulus clouds in the tropics. It was a presentation that captured my interest and admiration somewhat beyond the questions I asked about her work. It began professional communications between us that later were infused with more personal communications that led to a surprising scientific collaboration and—to a host of our colleagues and friends—an even more surprising and enduring marriage of 45 years.

Chapter 4
Early Hurricane Studies

Grady Norton and the Miami Hurricane Forecast Office (1944–46)

Housed in the penthouse of the 19-story Congress building in the heart of downtown Miami, and less than two blocks from the shores of Biscayne Bay, was the regional hurricane forecast office presided over by Grady Norton, a giant in the annals of the hurricane warning services of the nation. A giant also in his six-foot-two stature, yet he had a kindly and welcoming manner. In the perception of his radio audience, his voice was readily accepted as a mandate for emergency action: evacuate or find safe shelter! In gentler weather he was the very icon of a stately, erudite, but humble, Southern gentleman—proud but without an arrogant bone in his body. Interestingly, in both manner and appearance, he was almost diametrically the opposite of Isaac Cline (of Galveston, Texas, hurricane fame), both of whom I knew and respected well, but for different reasons.

This was the Miami and professional venue into which I was thrust, in May 1944, as one of two journeymen hurricane forecasters with a supporting staff under Norton. In that small office, we had four "subprofessionals" to do map plotting and to help with communications and some analyses. We mapped a substantial area from southern Canada to Panama, and the central Atlantic to the eastern Pacific, and constructed prognostic charts as necessary for our forecast responsibilities. I would say there was never a group I worked with more dedicated to serving the public completely than at that office.

Grady Norton was not only a well-known personality within his constituency, but he was also one of the most respected forecasters I've ever known, probably the most effective hurricane forecaster. However, it took me nearly two years of working with that man to realize that he NEVER issued a hurricane forecast if he didn't have to. He advised the public very astutely about what was out there, where the hurricane was, its previous path, the direction of movement, how hazardous and how important it might become...and the people listened! Listened with rapt attention, and when the storm got close and it was time to move, he said, "Run for it," and they ran! They listened, and they trusted him. He was a most effective public servant, but he never issued a forecast until the chips were down and the warnings had to go up and then he would call his shots clearly.

On the other hand, within the office, he was forever making forecasts and very good ones, often written on the margins of the analyzed weather map. During the hurricane of October 1944, he made a forecast for internal use while the hurricane was moving slowly westward in the Caribbean near Grand Cayman, calling for the center to move across Tampa Bay at midnight two days later. It did just that at 10 p.m., instead of midnight, one of the best hurricane forecasts I've ever encountered.

Afterward, I asked Grady, "How did you do it? You must explain the process you go through in making such a forecast."

"Well," he replied, "I just look at the steering currents and reason how they're going to change and make my forecast accordingly."

"But Grady," I continued, "I draw those streamline charts too, and I know you consider the wind at the top of the hurricane steers it. But I don't know how you can come up with the reasoning for changes in the circulations during the 48-hour period." After a little silence, he replied, "Well, if you really want to know, Bob, when I'm in doubt about something important like that, after thinking it over a long time, I go out on the penthouse roof, put my foot on the parapet, look out over the Everglades and say a little prayer. Then, when I return to the office, I know what I'm going to forecast."

Life in Miami

Peggy was only six years old when we arrived in Miami in late spring of 1944 and Lynn was still in her first year. We were fortunate to rent a comfortable (but elderly) residence at 1717 SW 11th Street, not far from my work at the hurricane forecast office on Second Avenue. Mazie had her hands full with two very active pre-school children while I was spending 10–12-hour days (Monday through Saturday, wartime weeks) with Grady Norton at the hurricane office, probing the things we DIDN'T know about hurricanes. When a hurricane showed up, the two of us worked 12-hour forecast shifts, sleeping on an army cot in a storeroom when the workload permitted. The head of the Florida Fruit Frost Service, Warren Johnson, was detailed to help out during hurricane emergencies. Suffice it to say, hurricane-free Sundays were about the only times the Simpson family was alone together, excluding the hours we spent in church services.

In retrospect, it is now difficult to grasp how much had taken place during the three years I was officially assigned to Miami. Under the tutelage and inspiration of Grady Norton, and the motivation he imparted to me to become involved in hurricane research, I burned the midnight oil applying what I had learned at Chicago, digesting the technical concepts and procedures for hurricane forecasting advocated by Norton, and developing research plans for testing these concepts. I was stimulated primarily by Grady's concepts about hurricanes but which, from my theoretical reasoning and subsequent analytical efforts, began to lose credibility with me. Norton's simplistic notion of hurricane structure was that the warm core of the hurricane would diminish the strength of its destructive winds with altitude, the low pressure center gradually disappearing, and thus the wind direction at the top of the storm would reflect the steering currents driving the entire system. While Norton had been very successful in applying this idea in his forecasts, it was inconsistent with theoretical reasoning and computations published in the early 1930s. This led me to consider exploratory research of tropical cyclones using reconnaissance aircraft that might get at the truth.

Panama

In 1945, there was an interruption in my Miami assignment. The Air Force asked Dr. Reichelderfer for the loan of Dr. Robert Fletcher and myself to help establish the School of Tropical Meteorology in Panama. This was a program to provide an educational retreat for A-course graduates who had been taught temperate latitude meteorology at Chicago or MIT but found themselves ill equipped to deal with the tropical weather. This program was similar to the Navy's Institute of Tropical Meteorology (ITM) in Puerto Rico that was established several years earlier. The Air Force school was at Howard Air Force Base, which had been carved out of the jungles in Panama. The training at the school, six days a week, was somewhat unique and quite intensive. Courses in dynamic meteorology were taught by Bob Fletcher, radar meteorology by Myron Ligda, and tropical climatology by Lt. William Odom.

I was responsible for a program on synoptic meteorology of the tropics, which combined a series of lectures along with laboratory exercises that made use of two C-47 transport planes (fondly known as gooney birds). These were modified to provide 11 training positions, each with navigation equipment, a free-air temperature readout, and a radar repeater scope at each position. The students were awakened three mornings a week at 3 a.m. and brought to the weather office, where they would plot and analyze the latest weather observations and then prepare a prognostic chart and forecast. Afterward, they would board a gooney bird and verify their route forecasts. One plane would fly north to Jamaica, the other would fly south to Salinas, Ecuador. When the planes returned, the students returned to the weather office (before dinner) for a postmortem discussion of the reasons for the weather encountered, and of the faulty reasoning, which made all too many of the forecasts fail. These laboratory exercises provided the focus for the lecture of the following day. To some students this was a rude awaking to reality, but it proved, nevertheless, an effective learning experience for all of us.

There were few textbooks on tropical meteorology in those days, so we used a set of notes and illustrations prepared by Fletcher and myself. It was not exactly a case of the blind leading the blind, although ignorance did abound. My experience at Chicago had brought me in contact with such people as George Cressman and Herbert Riehl who, from their work at ITM in Puerto Rico, had set forth some fairly sound and fundamental principals of tropical weather systems, including the concepts of the intertropical convergence zone (ITCZ) and of easterly waves. At Panama, of course, the ITCZ received at least as much attention as easterly waves. In those days we had five radiosonde observations in that area, some of which, including that at Clipperton Island (10°N, 110°W), were classified and could not be transmitted elsewhere in a timely manner. With these soundings we were able to make very good structural analyses of the ITCZ as it surged northward across Panama and then reformed to the south. Fletcher and I published several papers based on these flights.

In July 1945, when our gooney birds had proven their flight worthiness, we decided to reconnoiter the first hurricane that came within our flight range. A small tropical cyclone that was observed approaching the Lesser Antilles late that month turned out to be just what we were looking for. After predicting that it would move to a position in the central Caribbean just south of Hispaniola 48 hours later, we decided to deploy to Curaçao, overnight there, and launch our research mission from there the following day. It was a relatively audacious, if not injudicious undertaking, considering that operational flying in hurricanes had been initiated only a couple of years earlier when Joseph Duckworth, in 1943, had made the first deliberate flight into a hurricane eye. However, our plans worked out just fine.

The hurricane was located only a degree west and a bit north of the predicted position. The plan was to circumnavigate the center, keeping the wind on our port quarter, thus spiraling gradually into the eye. While at a flight level of 5,000 feet, we got only infrequent glimpses of the sea surface through the drift meter; the radar kept us in excellent touch with the eye center. The only problem, as it turned out, was that we found ourselves in heavy rain that was tucked into a corridor equidistant from the eye and the 8,000-foot mountains of Haiti's southern peninsula when the radar quit on us. Since we were already pretty well committed, we kept on the projected spiral path and wound up in the eye with little difficulty, entering from the southwest. However, we had used more fuel than expected, and did not have enough for a safe return to Curaçao. So, we declared an emergency and landed at Ciudad Trujillo (now known as Santo Domingo), much to the distress of the U.S. ambassador, since our relations with that country were a bit too tenuous at that time for a military plane to be landing there. He chewed out the unfortunate Air Force pilot quite eloquently when we landed. But all ended well, as we returned to Panama without further incident, and with a bevy of new questions to explore about hurricane structure.

On August 14, a few of us were in one of the gooney birds flying westward along the equator several hundred miles beyond the Galapagos when the radio announced Victory over Japan (VJ) Day. Several of us celebrated by dancing some kind of a jig along the narrow aisles of the gooney. Shortly afterward, the mission was called off and we headed back to the islands to join the merriment there.

The school in Panama was short lived after the war's end. While only three classes, six weeks each, passed through the school, serendipity had played a role in our experiences there, opening up new ideas about future use of aircraft for valuable atmospheric research. By the end of my nine months in Panama, I had begun two research papers on ITCZ structure. I returned to Miami in November 1945, and divided my waking hours between hurricane operations and research involvement. I managed to complete two more papers for publication on hurricane motion during my time at the hurricane warning center.

Washington, D.C.

I wanted to get on with my research, but I didn't feel it could be done effectively at Miami while dividing my time with forecasting obligations. I asked for and received a temporary assignment to Washington, D.C., during the winter of 1946. It ended up becoming a "permanent" transfer there early in 1947. At the outset my new assignment in Washington, D.C., was in the training branch, working with Al Carlin to provide retraining of military forecasters transferred to the Weather Bureau after the war. This included brief refresher courses in physics and dynamic meteorology as well as procedures and policies of the Weather Bureau in preparation for assignments as forecasters and in other professional positions. After several months with the training branch, I was transferred to a position as executive assistant to the deputy chief for Operations, Mr. Delbert Little.

Once my appointment turned permanent, I retuned to Miami and moved the family into a small upstairs apartment in Falls Church, Virginia, the only accommodation available on short notice. However, some months later we were able to find a fine accommodation in a two-story home in West Hyattsville, Maryland, where both girls enjoyed a memorable snowstorm in early January 1947—17 inches of snow virtually shut down most traffic lanes. I put chains on the car's tires, tied a rope from the sled I had just bought the girls to the car, and towed them around the residential area without fear of traffic—great sport and we all enjoyed the fun. Also, the family became active members and enjoyed the fellowship of Mount Vernon Place Methodist Church in the city, where we made many friends.

I still had to "bootleg" time to do research and research planning. By happenstance the office I shared with Delbert Little was directly across from the chief on the third floor, the "head shed" as it was known. Most of the staff members in those days were suburbanite carpool commuters who rushed away promptly at 5 pm. All but the chief, who customarily spent an extra hour at his desk catching up on correspondence and other matters that could not be accommodated during work hours, and myself. I usually did not depart until about 6:30 p.m., reading or working on one or another research subjects I couldn't address during the workday. After ducking into my room to see what was keeping me late at my desk, it became a habit for Reichelderfer to spend a little time at my desk to inquire how the research was going, often asking penetrating questions or contributing interesting suggestions. It was largely through these informal brief discussions that the chief helped me establish the contacts with Air Force hurricane reconnaissance groups, which led to my piggyback hurricane research missions.

The first "piggyback" missions

During the summer of 1947, I was able to arrange to spend six weeks at Kindley Air Force Base in Bermuda, where I had been given permission to ride piggyback missions with the Air Force's 53rd Weather Reconnaissance Squadron, the famous hurricane hunters, during hurricane reconnaissance flights. On these missions, once the operational information was obtained, I was allowed to use the remaining airtime to obtain research data.

The Air Force soon appreciated the potential value of these research data, and when the great hurricane of September 1947 came along (code named "George" by the hurricane hunters), I was allowed to mount a number of special flights devoted exclusively to research, using new WB-29s just being delivered to the 53rd squadron for hurricane recon from their Bermuda base. These were the first aircraft, I believe, to have analogue recorders for the meteorological probe systems. The main objectives of the proposed special research flights were to find out what really was going on at the top of hurricanes, especially the character of circulation there. Recall that Grady Norton was convinced that the circulation disappeared somewhere below the 300–200-millibar level and that the geostropic flow at that level comprised the steering current. In view of the earlier work of Bernhard Haurwitz, I did not see how the vortex could completely disappear at such a low level, but it was important to learn the distribution of vorticity at upper levels and to learn if the circulation near the top of the hurricane did relate discernibly to the steering of the vortex. So, I proposed to use one of the new B-29s to get as high as we could to try to describe the circulation near the top of this large hurricane.

After it had moved north of Puerto Rico, it passed over the Bahamas, across Florida into the Gulf of Mexico, and finally inland near New Orleans, first disabling the Miami hurricane forecast office and then briefly the New Orleans office. On two occasions—as it approached the Bahamas and, later, in the eastern Gulf of Mexico—high-altitude research flights were made, circling the center just outside the eye. With the superchargers installed on these new B-29s, we were able to reach a pressure altitude of about 36,000 feet, which in the warm core of this hurricane was close to 40,000 feet above sea level. Here, we skimmed through the tops of clouds as well as circumnavigated the eye four consecutive times. With five good loran fixes made during each circuit, it was possible to measure a mean tailwind component of 100 miles per hour, only a bit less than the maximum surface wind. As far as I know, this was the first observational evidence of the massive vertical transport of momentum by convective clouds in the eyewall; a process that created an imbalance between pressure gradient forces that diminished with height, and the angular momentum that remained almost constant, thus establishing the centrifugal acceleration needed to carry the warm air near the center out into the colder environment and with it a shield of outflow cloudiness. Of no small additional interest is that, a decade later, these observations foreshadowed the classical theoretical work of Joanne Malkus and Herbert Riehl regarding mesoscale vertical mass transport of angular momentum in hurricanes.

I returned to Washington, D.C., convinced this method of exploring hurricanes had demonstrated its importance as a tool of severe storm research and understanding, provided the aircraft could be equipped with suitable digital data recorders. The analog systems we were using gave apparently useful information on gradients of values, but the absolute values of the data obtained, especially of temperature, were not taken too seriously. My Washington assignment in those days was primarily in operations, not in research. However, since my desk in the central office was just outside Dr. Reichelderfer's office, I had the opportunity to spend many after work hours reviewing with him and Dr. Harry Wexler, director of research, the data obtained from my research flights. I not only received much encouragement but also many excellent suggestions from both. A number of descriptive papers did come from this effort. Reichelderfer really felt the need for research in tropical meteorology in those days and asked me to develop a plan and budget to systematically extend the hurricane research effort. The first plan was submitted to Congress in 1948…and was quickly rejected.

Hawaii and the Pacific Region

What an extraordinary and challenging sequence of venues I had had in 13 years! Beginning with Crockett in 1935, I had traveled to Fort Stockton, then Corpus Christi as a music teacher, then as a meteorologist from Brownsville in 1940, to Swan Island, New Orleans, Chicago, Miami, Panama, back to Miami, and finally Washington, D.C. For me it was continually exhilarating, often exciting. But it proved difficult to share the exhilaration with the rest of my family, who often had to be left behind for months at a time between changes of assignment. This strain on the family bonds proved too much to bear in early 1948, when Dr. Reichelderfer asked me to accept a four-year assignment to Hawaii. This would turn my back on all the exciting exposures and new research concepts about hurricanes and tropical meteorology I had planned to pursue from the Washington, D.C., area. My wife, Mazie, adamantly opposed the move and in consideration of what all these moves had cost my family, I turned down Reichelderfer's request. However, due to the urgency with which he pushed this opportunity to advance my career, I reversed my decision. My wife was adamant. After further discussions with Mazie, I asked for a formal separation to which she agreed, with the understanding that the option to reconsider separation could be sought by either of us within a reasonable time. As mutually agreed, in late April 1948 I returned Mazie and the two girls to Texas, resettling them and agreeing on items of support and other necessities, and then continued on to Hawaii.

Despite what I considered an adequate emersion in available literature about Hawaii, both social and scientific, I was still unprepared for the ceremony, glamour, and indeed the mystique of Hawaii. As the beautifully refurbished and fully air-conditioned SS *Lurline*, the proud flagship of the Matson Lines, gently docked before Honolulu's Aloha Tower, I was welcomed by the enchanting strains of "Aloha 'Oe" by a Hawaiian chorus and band. I soon discovered a dozen or more Weather Bureau families, who had spotted me at the rail of the main deck and were waving their traditional leis of welcome. It was a never-to-be forgotten experience. However, within a week of my arrival, I was nonplussed to discover that almost all the families who greeted me so welcomingly at the Honolulu Harbor had been motivated at least in part by the hope that a new boss, especially one from Washington, D.C., would help them obtain a reassignment back to the mainland. Why on earth would anyone assigned to such a paradise be so eager to leave it? In little more than a week, which included a few interviews with those seeking a means of getting back home, and after my first motor trip encircling Oahu, I realized that the extensive low morale came from the realization not just how far from home they were but the discovery that it was too costly to transport a family back home for a vacation. Aha!—my first but immediate challenge. At the time I didn't realize how easy it would be to deal with this problem, by indirection rather than a frontal attack.

On my reassignment papers, my new responsibilities were 1) to negotiate and supervise the transfer and management of all weather stations established by the U.S. military in the Pacific basin during World War II to the Weather Bureau; 2) to consolidate the four existing Weather Bureau offices in Hawaii (aviation, local weather services, agricultural services for all Hawaii, and climatology); and 3) to upgrade the weather service facilities for the Philippines and Trust Territories, all under the oversight and administrative management of a single office in Honolulu. It was a tall order for my first assignment, even with sole directorial responsibilities, answering only to the chief of the Weather Bureau. Fortunately, however, the transfer of weather services from military to civilian responsibility had to be co-ordinated by the Air Coordinating Committee (ACC), chaired by the Navy under the command of Adm. William Radford, comander in chief, Pacific command (CINCPAC). By default, the Weather Bureau member of this committee became a member of Radford's staff with access to most naval officer's social facilities at Pearl Harbor, an advantage that promptly resolved the morale problem among Weather Bureau personnel. I ensured that monthly Weather Bureau parties were scheduled at the Navy Officer's Club at Pearl Harbor. Also, there were privileges of much greater significance and lasting value to science that evolved in connection with the ACC membership, developments detailed in Forrest Mims's recent book on the exploration of Mauna Loa and the development of its observatory, *Hawaii's Mauna Loa Observatory: Fifty Years of Monitoring the Atmosphere*, summarized briefly in a later chapter.

While in Hawaii, my interest and concerns with tropical meteorology, and the research objectives generated in the Atlantic basin remained as keen as ever. In fact several of the more significant scientific papers of my career were products of research conducted there. One area that provided rewarding research was my study of Kona lows. These were small, low pressure cells that formed near the archipelago near the southern end of the subtropical jet, which often carried moisture onto the dry, or kona, side of the islands. Neither tropical nor extratropical in nature, they had been understudied by meteorologists. I demonstrated that they contributed as much as two-thirds of the rainfall that fell onto these dry sides.

My administrative responsibilities included far-reaching travel to Japan, Australia, New Zealand, and islands of both the North and South Pacific, as well as official involvements in a wide variety of scientific agencies and organizations in Hawaii, including the Hawaii Academy of Science, the Bernice Pauahi Bishop Museum, the Pacific Science Congress, the Pineapple Research Institute, the University of Hawai'i at Mānoa, the Federal Business Association, and the U.S. Geological Survey. Fortunately, I was blessed with an eager and competent supporting staff in my office in downtown Honolulu, especially from Art Youmans, my administrative assistant and office manager; Helen Tsuschia; and Miss Chen; and at the forecast office at John Rodgers Airport, by Charlie Woffinden and Jess Gulick, who relieved me of nearly all technical and routine administrative tasks within bureau and office policy. At the airport I kept a private office and a small library with a decade of historical weather maps. Here, I was able to do most of my research without disruption save for a telephone with a private line.

Only a few weeks after my arrival, Moku'aweoweo, the volcano at the summit of Mauna Loa, erupted, sending a smokestack of lava plumes high above the crater rim. The nighttime glare was clearly visible at night in Honolulu, more than 200 miles away, a unique display posing a number of interesting questions beyond that of visibility at that distance. My primary concern was the vertical profile of atmospheric heat dispersion above the volcano rim. At my request, the Hickam Air Force Base dispatched a fighter plane with recording external temperature probes to measure the changes in vertical profiles of temperature in successive diametric passes above the caldera up to 20,000 feet—a record. I failed to find useful data for the purposes I intended; however, during the Pacific Science Congress I attended in New Zealand soon after, I captured the interest of the British scientist Sir Geoffrey Ingram Taylor, in the paper I presented about these data. Later, in a letter, he expressed appreciation for the data I had supplied and told of his ongoing research using the data.

I spent two weeks in New Zealand, one week in Auckland on the North Island, the second week in Christchurch on the South Island. The long weekends between and following the two sessions featured bus tours to areas of particular interest to the scientists. Then, Dr. Reichelderfer authorized an extension of my travel to Australia to report on research underway in Melbourne, and for discussions with Edward George "Taffy" Bowen on the latest developments in cloud seeding research in Sydney.

I was pleased, excited in fact, with this additional opportunity. However, I was less than impressed with the research efforts in Melbourne, and found Taffy Bowen's weather modification program in Sydney curiously primitive, but inventive and persuasive. Taffy arranged a cloud seeding demonstration flight in his old Dakota seeding plane (Australian version of the DC-3). The "seeding" was done by removing a window in the plane's cabin through which crushed ice was shoveled out into the cloud tops (quite an experience, though somewhat lacking in observable responses).

Overall, the lengthy trip (nearly six weeks) provided me valuable acquaintances in many areas of the Pacific as well as broader scientific acquaintance with weather problems across the area. Also, I returned to Honolulu with new international responsibilities, having been named executive secretary of the meteorology section of the Pacific Science Association under its chairman Athelstan Spilhaus, who delegated most of his responsibilities to me.

Under the aegis of the ACC, I made six very useful and enjoyable business trips to Tokyo, Japan, during the days of Japanese reconstruction under the guidance of General Douglas McArthur. On a more personal level, these trips cemented a rewarding scientific and personal relationship with (then) Col. Tommy Moorman, commander of the Air Weather Service 2143rd Air Weather Wing in Tokyo, and chair of the meteorological branch of ACC. This would lead to his support of my research flight into Typhoon Marge in a couple of years and later to Air Force support of the National Hurricane Research Project.

A Reunion, Then a Divorce

Several months into my Honolulu assignment, Mazie wrote that she wanted to reconsider the separation and was prepared to come to Hawaii in a week or two. So, I scrambled to find a rental home and acquired one in the Mānoa valley near the university campus and the school Peggy would attend. It was so good to have Peg and Lynn back, and I immediately told them of the wonderful things in store for them in Hawaii. And Mazie and I, with tongue in cheek and a forced note of happiness, set out to try to restore and improve what had been set aside. However, within a month it became clear to us that it was a futile undertaking. We agreed to seek a divorce. I bought their plane tickets for returning to San Antonio, with a stopover in Los Angeles. I telephoned George Kalstrom, a classmate from Chicago and a close friend, who was in charge of the forecast office in Los Angeles. He readily agreed to have Mazie and the kids met at the airport, accommodate them overnight, and return them to their flight to San Antonio the next day.

With this parting, I can't recall another exchange of communications with Mazie, except terse ones to coordinate meetings or visits with Peg and Lynn. But I exchanged frequent cards and letters with my daughters to learn of their activities and interests, and to describe interesting encounters and activities during my travels, about which they both seemed fascinated to learn. And rarely did I travel overseas without sending each of them some small gifts or mementos of that travel.

The Mauna Loa Experience

Of all my experiences encountered while directing Pacific operations, none was comparable either in immediate personal gratification or ultimate importance to science than the establishment of the Mauna Loa summit observatory. The ultimate success in establishing a permanent observatory near the 11,000-foot level led to the long-term record of increasing atmospheric carbon dioxide that generated international concern about global warming. The account of the local cooperative support in establishing an initial foot-in-the-door structure at Mauna Loa's summit and the follow-on collaboration with Dr. Ralph Stair of the National Bureau of Standards to create the permanent observatory is well told in Forrest Mim's book titled *Hawaii's Mauna Loa Observatory: Fifty Years of Monitoring the Atmosphere*. A summary of my cooperative efforts with this venture is in the following chapter.

Soon after my arrival in Honolulu, Dr. Luna Leopold, a distinguished geophysicist and a water resources expert with the Pineapple Research Institute (PRI), introduced himself and offered me, on behalf of PRI, to join in an exploration of principal areas of Hawaii of mutual concern. The exploration carried us to the more remote and less accessible regions of Kauai, Oahu, Maui, and Hawaii Island. This gave me an acquaintance and appreciation of the distinctive physical features, the wildlife, and agriculture of each island. But despite the wonderment and individual uniqueness of each island, the eyes of this meteorologist were captivated by Mauna Loa. Its ever-so-gentle slopes extending upward almost to the midtroposphere, far above the prevailing trade wind temperature inversion and the day-to-day weather changes, clearly made it the most promising venue for meteorological studies. The inversion trapped heavy tropical rains below while maintaining above it a dry landscape. The arid air promised minimal turbulent mixing and the flow of pristine air thousands of feet above the vegetated area, air that reached Hawaii after thousands of miles of normal oceanic transport downstream from the nearest sources of pollution. Where else in the world could you find so idyllic a site for atmospheric observations and experimentation?

My interview with a science reporter from the *Honolulu Advertiser* stirred public and private interests in support for establishing an observatory atop Mauna Loa. Less than two weeks later, Tom Vance, director of Hawaii institutions, appeared at my door with a proposal. "I think we might be able to do business together in achieving our common interests in Mauna Loa. I have a prison camp at Kulani deep in the rain forest at the 5,000-feet level on Mauna Loa. The inmates there need a challenge to help equip them to be responsible citizens when they leave prison. They can help both in road building and in construction of the observatory building." Not until later did he confide that what he really wanted to do was establish a ski lodge at the summit to be operated by the inmates.

This at first seemed a questionable way at best to go about the establishment of the observatory, but after further thought and consultation with Dr. Reichelderfer, I decided to pursue the matter further. The big question was how to acquire the road building equipment to construct the access road. To explore the options, I sought out my colleagues at the Joint Meteorological Center (Pacific) at Pearl Harbor. I explained what we wanted to do and told them, "I can get the help of the Geological Survey at the volcano observatory to help stake out a roadway likely to be safe from lava flows, and can provide from my regional funds the cash to purchase building materials and equipment for the observatory. The Territory of Hawaii can supply the labor to construct the building and to build the road. However, we don't have the equipment to build the access road. Can you help us?" Admiral Radford agreed to lend the project two big road building Cats and to ship them to Hilo, providing others would pick up the cost of fuel and spare parts. I quickly agreed that the Weather Bureau would do so. The project was set in motion, supported officially by only a handshake or two.

It was a small but auspicious beginning. The real fight came, however, after getting final approval from Dr. Reichelderfer, in persuading the Secretary of the Interior Oscar Chapman to allow us to use national park land to build the road and to erect the observatory building. This was a tough job. But with Reichelderfer's help, we finally got approval over strong objections by Chapman's staff.

Plans were finally completed to put a small observatory at the summit of Mauna Loa, just below the caldera of Moku'aweoweo crater at an altitude of 13,453 feet.

The building was about 10 feet by 10 feet, with heat and a cooking facility, a bunk for overnight visits, and oxygen to supplement the thin air, in addition to the autographic equipment and probe systems to record wind, temperature, pressure, and rainfall. A support base was set up at the Hilo Weather Office with additional staff so an observer could visit the summit facility once a week to change recorded sheets and to check the equipment. Gordon MacDonald, director of the Hawaiian Volcano Observatory, made a number of trips to the summit with me and on one occasion with Charles Woffinden, head of the Weather Forecast Office at Honolulu, to stake out the access road. The building itself was constructed in sections at the Kulani prison site and transported to the summit on a big truck.

With the work complete on December 12, 1951, a big inaugural service was held at the summit site that was attended by some 50 officials, including Sam King, the governor of the Territory; Washington representatives from the departments of Commerce, Interior, and Navy; as well as a number of scientists from universities and government laboratories. All were transported to the summit in four-wheel-drive vehicles. The dedication by Governor King took place during a light snowfall and was followed by an open-pit barbecue around a blazing bonfire.

We were off to a humble, though significant, start that received wide coverage in the press. But it fell far short of what was ultimately hoped for in regard to ease of access or the type of measurements we made. The road left much to be desired. Despite the fact it had been carefully sealed with cinder cone material, it deteriorated rapidly. My plans were, ultimately, to have more comprehensive measurements requiring monitoring by personnel in residence at the observatory. I had, among other things, hoped to make daily measurements of total ozone.

Why ozone? Because I had been impressed by the potential significance of the cold lows that continually paraded westward from California across to Hawaii, many of them eventually reaching the western Pacific, where they seemed to be associated in some way with the formation of typhoons. Clarence Palmer had proposed an extraterrestrial source of energy for these lows, a very controversial idea. Nevertheless, it was my idea that we might learn a great deal about these lows and how they might ultimately contribute to typhoon formation if we could study them as they moved over Hawaii, using ozone as a tracer. This and other, perhaps more sophisticated, equipment would have to await more substantial funding; but we did have our foot in the door.

After I left Hawaii, however, it got progressively tougher to maintain the facility at the summit site. Finally, the access road deteriorated to the point that it was decided, much to my distress, to close the observatory altogether. To me, we had established a foothold at a site with enormous potential for scientific observations of many kinds, and I felt we had been less than perceptive in tossing in the towel so soon.

Typhoon Marge

In 1951, at the request of the U.S. Weather Bureau, Col. Thomas Moorman arranged for a specially instrumented RB-29 aircraft of the 2143rd Air Weather Wing to support plans for a historic typhoon research mission as an extension of the squadron's regular recon responsibilities. The aircraft would be sent on a regularly scheduled reconnaissance mission, but between the required center "fixes," Weather Bureau personnel would be at liberty to use the plane to investigate their research interests. The additional instrumentation would allow us to record various meteorological aspects of the storm that had previously been unexamined.

On August 15, Typhoon Marge was 950 miles northwest of Guam and threatening Taiwan. I deployed to Guam to participate in a research flight aboard the RB-29. Details about the flight were published in an article in the *Bulletin of the American Meteorological Society*.

> We took off from Guam shortly after dawn... The plane climbed slowly to an altitude near 11,000 feet beneath a dull overcast of cirrostratus and altostratus clouds. Even at the distance, some 700 (statue) miles from the storm center, the overcast apparently was a part of the great shield of high clouds associated with Marge...
>
> There was an ominous absence of isolated or random convective cloudiness. With the exception of several well-organized lines of towering cumulus encountered periodically, there were only patches of small stratocumulus and stratus clouds beneath the plane... Proceeding toward the storm center, the base of the altostratus shield lowered steadily until it finally engulfed the plane approximately 150 miles from the eye. From that point on, only occasional glimpses could be had of the sea surface. Navigation toward the eye thereafter depended on proper interpretation of the squall-line pattern visible through radar, and of pressure variations.
>
> Continuing on instruments, radar soon detected the edge of the rainless typhoon eye directly ahead. A series of heavy rainbursts followed and then several bumps of moderate turbulence as the plane broke through the walls of the eye into clear air.
>
> Here was one of nature's most spectacular displays. Marge's eye was a vast coliseum of clouds, 40 miles in diameter, whose walls rose like

galleries in a great opera house to a height of approximately 35,000 feet, where the upper rim of the clouds was smoothly rounded off against a background of deep blue sky. The sea surface was obscured by a stratocumulus undercast except for two circular openings on the east and west sides of the eye, respectively. Clouds in the undercast layer were grouped in bands that spiraled cyclonically about each of these openings, or clear spots, both of which were approximately five miles wide. This horizontal alignment of clouds suggested the possibility that two separate small eddy circulations were present within the eye envelope...

From the weather observer's position in the nose of the plane, where a view almost vertically downward can be had, the sea surface could be seen occasionally through breaks in clouds. As the plane skirted close to the north wall of the eye at an elevation of 8,000 feet, fleeting glimpses of the seas below revealed an amazing state of turmoil. Waves were scarcely distinguishable here and the ocean was almost completely obscured by the mad rush of greenish-streaked white froth. Most of the crew were veterans of many typhoon reconnaissances; however, none had ever seen a sea condition, either inside or outside a typhoon eye, which compared with that glimpsed from this vantage point...

With fuel growing short, the plane upon completing the second ascent in the eye departed for home at an elevation of 17,000 feet. As the walls of the eye were reached, moderate turbulence was felt, somewhat greater and more continuous than had been felt when entering the eye through the same quadrant at 9,000 to 10,000 feet...

During the mission I took many photographs of the storm, its environment, and most especially its spectacular eye. The unprecedented pictures of the eyewall exhibited downward-sloping vortices embedded in the clouds. And I documented the existence of a low "hub" cloud that rode in the center of the eye, which suggested the low-level circulation in the eye converged toward a central circulation point. This became an important point in later discussions concerning hurricane beacon balloons.

The research objectives of this flight into Typhoon Marge as planned were ambitious, challenging, and unique for aircraft exploration at that time. But when these objectives were completely fulfilled, the findings validated, and compared with earlier records, it was clear that the mission had been of historical distinctiveness. Marge, at the time of this flight, had been the most intense typhoon of record to be fully explored by aircraft. Marge's central pressure, measured by dropsonde, was 891 millibars, one of the lowest of record at that time. Finally, during the flight's two ascents into the eye, one to a height of 27,000 feet, the air temperatures remained virtually constant from near the surface to 500 millibars (the middle troposphere). This exploration offered new insights on tropical cyclone eye structure and dynamics, and further supported the need for aircraft reconnaissance for tropical cyclone research.

Kate Smith

My years in Hawaii were mostly consumed with professional activities both in Honolulu and in travel across the Pacific. Social activities were confined almost entirely to those intended to improve the esprit de corps of Weather Bureau employees and their families eager to return to the mainland. So, after the departure of Mazie, and the divorce that followed, I moved into a small apartment in the Waikiki sector of Honolulu near the waterfront, facing the Ala Wai Canal, and within walking distance of my downtown office. Despite the social activities, primarily sponsored for or by the Weather Bureau employees and their families, and often weary from the stresses of daily professional obligations, I soon found myself becoming a very lonely man.

When in Honolulu, I usually ate alone at a small cafe on Waikiki near my apartment. One early evening, however, I arrived to find the last table being taken by a young lady I hadn't met but occasionally passed in the hallway of my apartment en route to work. Recognizing the crowded conditions, she invited me to share her table. I did and in the course of the meal we became better acquainted. Her name was Kate Smith, and it turned out she was the office manager and assistant to the vice president of Hawaiian Airlines, a man by the odd name of Ford Studebaker with whom I had recently become acquainted in connection with sponsorship and support of the Mauna Loa summit observatory, and also in my capacity as president of the Federal Business Association.

Kate and I soon became close personal friends, enjoying each other's company, sharing the same apartment and the relatively small amount of free time together that I had in Honolulu. But unfortunately these rarely extended to social events involving the Weather Bureau families with whom she remained essentially estranged, never acquiring an interest in my involvement in meteorology or related associations in science, or the travels they involved. Nor did I find the time to become involved with the few social friends she had cultivated in Hawaii, other than those she was involved with in her work at Hawaiian Airlines. Long before I returned to Washington, D.C., it became evident to us both that the mutual attraction brought about by loneliness no longer existed. Yet, soon after I returned to Washington, she followed me there, where it took only a few months to see the relationship could not, and should not, continue to survive. It didn't. Distasteful and embarrassing as this career disruption was, it left few personal lingering scars. Some may have wondered why it didn't, but I was too busy involved and fascinated with career objectives to deal more decisively with this failure. After the day of our parting, I never again encountered or heard from Kate personally or indirectly.

Chapter 5
Weather Bureau Headquarters

Reichelderfer's Castle

After exactly four years, in precise accordance with Reichelderfer's promise, I left Hawaii after a gracious aloha party by Honolulu friends and associates. Once again I stood before the Aloha Tower and boarded Matson's SS *Lurline*. I returned to Washington, D.C., to my same old desk, but with a handsome promotion and a much more attractive job description than I'd had when I left for Hawaii. However, I was leaving a venue that had afforded almost boundless freedom to pursue a career in meteorology, which had offered diverse challenges and was now a faraway land of enchantment, in order to return to an uncertain but more regimented future.

The Weather Bureau central office was at 24th and M Streets NW in Washington, D.C., in a four-story modern brick building fronting on M Street. The site was originally intended as a campus of the castlelike Spanish embassy. The main building and its surrounding structures, including stables and a string of utility buildings, had reverted to the U.S. government, and had been reconditioned for the use by the Weather Bureau. They served as quarters for various bureau functions, including a small auditorium for seminars, a print shop that produced the daily weather maps, and the Research Division.

The new construction was completed shortly after Reichelderfer was appointed chief of the bureau in 1938. A Roosevelt appointee, he came to the bureau following a distinguished naval career as an aerologist and a specialist in weather prediction for lighter-than-air operations, and had extensive graduate-level training in modern concepts of air mass and frontal analysis, introduced by the Norwegian School of Meteorology during World War I. But these theories had virtually been ignored and had never been applied in the civilian weather service of the United States, which Reichelderfer intended to change.

From this building flowed various centrally produced weather service guidance, including a daily weather map for the United States. It contained two separate extended weather forecast facilities, one headed by Charles L. Mitchell of the older school of prediction and the other by Jerome Namias, an MIT graduate employing the revolutionary concepts of Carl Gustav Rossby. Rossby also had been a graduate of the Norwegian school and had revised and upgraded their concepts by introducing three-dimensional understanding available from the upper-level sounding balloons introduced in the 1930s. Rossby's initial employment at the Weather Bureau had been rough and he soon left to found the Meteorology Department at MIT. Reichelderfer had brought him back as part of his efforts to modernize the bureau's services. Rossby was to oversee the retooling of prediction. Of course, it should be recalled here that Rossby soon found reason to separate from the bureau on the eve of World War II to establish the new Institute of Meteorology at the University of Chicago.

Namias was hired at the beginning of World War II in an effort to extend the Weather Bureau's forecast limit out to an unheard-of five-day horizon. He headed the Extended Forecast Branch until his semiretirement to the Scripps Institution of Oceanography at the University of California, San Diego.

With a stroke of optimism, if not genius, Reichelderfer appointed Harry Wexler, another MIT product, to head the Research Division. These humble quarters in Washington, D.C., were considered by some to account for why Harry seemed to spend more time traveling than at his desk in the stables! I should quickly add that in my view, if this perception is valid (and I doubt it), his time was in all likelihood spent more productively than if he had holed up at his desk. Harry published many very significant scientific contributions during his career, a number of which were in collaboration with other distinguished scientists, a few of whom were employees of the Weather Bureau, many from other disciplines.

There were frequent exchanges of views among the researchers and the operational people in the central office, mostly as a result of the efforts of Wexler who, to his credit, sponsored weekly seminars in the auditorium and insisted that operations people as well as researchers attend. This brought together and evoked lively discussions among people with broad and diverse interests, not only from the Weather Bureau, but also the Air Force and Navy. In the process, hurricanes and the problems they posed received a great deal of attention.

Modernizing the U.S. Weather Bureau

However, Reichelderfer was less fortunate in delegating responsibility for implementing his modernization efforts. His impatience with delays he encountered from his staff led to an increasing reluctance on his part to delegate responsibilities to his staff, and with it a growing difficulty in implementing many critical policy decisions. As one who was close enough to observe the growth of these management problems, to me they were the outgrowth of his own quick-wittedness in evaluating solutions to issues and impatience with his staff members, who were often unable to follow, respond, and carry out his orders. It was common to see him storm out of a staff conference in disgust because of the lassitude of staff in picking up on his proposals or queries about alternatives. That said, my appreciation grew for that quick-wittedness, interest, queries about the after-hours work I stuck around to pursue, and his swift responses (not always favorable) to my suggestions and requests, some of which I will mention later because so many of them made a difference or sent my curiosity or thinking racing off on a different tack. In any event, this is the Washington environment to which I returned in 1952—little changed in my four-year absence.

The Texas Radar Network

Working out of the offices of Delbert Little, deputy chief of the bureau, my official occupation was taking on ad hoc tasks as Reichelderfer's troubleshooter. My first assignment was resolving a dispute in Texas regarding the issuance of tornado warnings. Following the significant loss of life and damage from two tornados in 1952, one in Waco and the other in San Angelo, the Texas governor assembled a council to consider ways of establishing an independent state facility for prediction and warning of tornados. The U.S. Weather Bureau had no such facility at that time. Reichelderfer asked his Navy counterpart, Captain Howard "Shorty" Orville, head of Naval Aerology, to attend the council meeting and explain the limitation within present scientific knowledge to predict the location and timing of tornado occurrence. Reichelderfer asked me to represent the Weather Bureau at the meeting and to seek a viable means of addressing the problem, hopefully within the jurisdiction and available resources of the federal government.

After a morning of testimony and discussion among council members, and a lengthy noontime conference call to Washington, D.C., summarizing the events of the morning, I was authorized to propose a cooperative plan for installing a network of surplus World War II radars modified to detect tornadoes as they formed, with a communications facility to connect all radar stations and to share information on tornado location and movement. This was providing that each city desiring to participate in the network contributed the funds to modify and install the radars. The proposal was promptly accepted by the council, and the Texas Radar Network was in operation in 1953 with such success that Oklahoma, Arkansas, and Louisiana applied to participate in this unique network. But they were refused. Instead, Congress in 1955 authorized the establishment of a national network of state-of-the-art (WB-57) radars to detect and track severe local weather systems.

My experience in Texas was a follow-on example of what I had done a few years earlier in the establishment of the Mauna Loa summit observatory, a carefully orchestrated effort of federal cooperation with the local agencies of a motivated community. It was the second of numerous such unexpected and unprogrammed challenges, opportunities found by facing unpromising situations and by finding unusual solutions that made a difference while adding zest to an already exciting life as well as a boost to the scientific outcome.

A Lucrative Offer

Back in 1952, when the private practice of meteorology was given great support by Dr. Reichelderfer, I had an uncle who was a well-known Texas banker, protégé of Jesse Holman Jones, and at that time president of the First National Bank of Commerce in Houston. He called me and said, "Bob, I think the time has come when Houston needs a more specialized weather service than we are getting from the Weather Bureau. If you will come down here and set up a private practice in meteorology, I will underwrite your initial and continuing costs for three years. Your business should be profitable by that time."

That was a shocker to me. First, I had never once contemplated such a thing. Second, I was just getting steamed up about the work I was doing for Dr. Reichelderfer and Delbert Little, all of which was giving me exciting new ideas about how to create a better Weather Bureau. I told my uncle I'd think about it and get back to him within a week to 10 days. I talked with a lot of people who were going into private meteorology at that time, one of whom was A. H. Glenn, who was in business in New Orleans at the time. Finally, I decided against the venture. Although I never doubted I could become successful in private meteorology, I had invested 12 years in government and looked forward to a career there. So, at the end of 10 days, I called my uncle back and said thanks but no thanks.

"Piggyback" hurricane research missions

The excitement and motivation generated by the over-the-top research flight into Hurricane George, staged from Bermuda in 1947, followed by my 13-hour three-dimensional flight into Typhoon Marge out of Guam in 1951, set the stage for a series of what were later know as piggyback research missions in the summers of 1952–54 from Bermuda. They were referred to as piggyback because they were superimposed on Air Force operational flights into hurricanes. The scientific experiments were conducted after the operational objectives (the location, strength, and direction of movement of the storm) had been obtained. The remaining flight time was then used to conduct carefully planned data acquisitions for research use.

The justification for these cooperative flights was the success of the George and Marge missions and the impact of the published papers following those flights. Reichelderfer had agreed I should spend summers in Bermuda explaining the research objectives to the Air Force hurricane hunters and developing flight plans that could achieve these research objectives: plans that were carefully rehearsed with all crew members, plans that could be directed by an onboard research meteorologist, and plans that could be adjusted during the flight as necessary depending on the operational information acquired during the flight. Once the operational objectives had been achieved, the research scientist would change seats with the operational flight director and occupy the forwardmost viewing position in the "greenhouse" nose bubble of the B-29 and B-50 aircraft.

In early September of 1953, Hurricane Dolly skirted the northern Caribbean islands, drenching Puerto Rico and the Virgin Islands with torrential rains. It then moved on a northwest course away from the islands. I arranged for a piggyback mission on a reconnaissance flight the morning of September 10, while Dolly was some 700 miles southwest of our Bermuda base.

I wrote about my Dolly mission in an issue of the *Bulletin of the American Meteorological Society*. Here are excerpts from that article:

> The flight from Bermuda into Dolly, whose center was several hundred miles east of the Bahama Islands, was at 7000 feet along a track which passed about 10 miles west of the eye, the subsequent penetration to the center being from the south. During the 12½ hours of flight, the aircraft was in the active storm vortex for more than 8 hours. Traversals of the eye and of the several storm quadrants were at 1500, 7000, and 27,000 feet. At 27,000 feet it was necessary to avoid the sector immediately in advance of the center because of icing, and because dropsonde signals could not be received from this area of massive nimbostratus. Dropsondes were made from elevations above 20,000 feet near the center, and at positions 60–100 miles west, south, and east of the center.... Loran positions

were obtained at frequent intervals and over-lapping space-mean winds computed to augment the multiple- or single-drift spot measurements of wind which were obtained wherever possible. In addition, numerous photographs were taken of the clouds in and near the eye....

..."Dolly" was not a large-size storm. Even though winds up to 100 mph had developed just northeast of its center, the growth of this storm had, for one reason or another, been stunted; it had failed to acquire circular symmetry of pressure fields even close to the center; the central pressure was little less than 990 mb and the familiar hurricane cloud system was well developed only in the forward quadrants of the storm. With less than two days after inception to develop and distribute its kinetic fury, moisture, and potential energies throughout the vortex, "Dolly" entered the recurvative cycle an immature hurricane....

...The structure of the hurricane vortex, even in the lower troposphere, is sufficiently complex that only limited conclusions can be drawn from an analysis of no more than two or three soundings from any given storm.

After our return to Kindley Air Force Base, Dolly took a northeast turn and aimed its 100-mph winds at Bermuda. The Air Force evacuated its aircraft to Tampa and further missions into Dolly were precluded. Before striking the island, however, Dolly weakened considerably and its effect on Bermuda was negligible. A week later, Hurricane Edna passed very near the island, wreaking considerable damage and bringing an end to any further piggyback missions for that season.

I returned to Washington, D.C., determined to propose funding of hurricane research using aircraft to probe these storms. My first proposal, developed in 1952, didn't get anywhere in that budget year, and neither did the one in 1953. I remember that in 1953, the annual budget of the Weather Bureau was just $27.5 million for all of its operations worldwide, including observations, forecasting, support services, and research. There was no extra money for new research initiatives. This all changed in 1954.

The hurricane season that year began quietly enough. In those days the Weather Bureau used the same list of women's names each year, so we again had storms Alice, Barbara, Carol, Dolly, and Edna. In late August, Hurricane Carol formed north of the Bahamas and brushed Cape Hatteras and then aimed for the Northeast. She smashed into Long Island and Connecticut, dumping torrential rains, bringing destruction not seen since the great New England hurricane of 1938. Carol's winds had not stopped howling when another disturbance formed off the Windward Islands. As it turned west while north of the Virgin Islands, it became a tropical storm and was named Edna. On September 8, it reached the Bahamian waters that had bred Carol just two weeks before and began to track northward following Carol's path.

By this time, I was in Bermuda, waiting with the Air Force's 53rd Weather Reconnaissance Squadron to participate in piggyback missions. On September 9, Edna, now a major hurricane, was moving slowly northward. I participated in a flight that day, but time allotted for observation was brief. The next day, as Edna picked up speed and brushed Cape Hatteras, I was able to fly in the storm for four hours, allowing me to make a complete set of observations of the hurricane eye and rainbands. On the first day, the eye was circular and over 20 miles in diameter, completely surrounded by magnificent nimbostratus, which towered to 30,000 feet. On September 10, the eye had expanded to an oval of 25–35 miles, and cirrus at very high levels moved in from the north and covered the storm core. The rear eyewall dissipated from the top down. Edna's faster forward speed was distorting it from its classic structure of the day before.

On both days, I observed inside the eye a mound of low-level stratocumulus clouds near the center that rose to about 8,000 feet. Near the eyewall, this mound tapered down and a number of breaks formed a moat between the mound and the eyewall through which the sea below could be seen. The shape reminded me of a hubcap, so I dubbed it the "hub" cloud. I had observed similar cloud formations in Typhoon Marge and Hurricane Dolly, and I and other scientists had begun to speculate on what caused these hub clouds to form and be maintained. I thought that in the atmospheric surface layers, the gradual frictional transport of air from the maximum winds under the eyewall progressively fed a relatively weak cyclonic circulation in the eye itself. These currents would converge at the wind center in the middle of the eye where they would rise, creating the hub cloud. I thought that if a properly designed constant-altitude balloon could be released from aircraft into the eye below the level of nondivergence, it would ride with the storm, remaining at or near the circulation center for many hours, possibly days. Attaching a radio beacon to such a balloon meant the storm center could be tracked remotely, even when no reconnaissance aircraft were present.

I was able to carry out several traverses of the spiral rainbands of Edna at altitudes from 7,000 to 12,000 feet. I tried to plot out temperature variations across the bands, but the variations were so small that errors that were due to the wet-bulb effect masked any trend. However, by comparing the pressure and radar altitudes (known as the D value), I could profile the pressure variations across the bands, and I noticed in some of the passes that the D value dipped below expected values on the interior side of the band. A careful analysis of the Edna and Dolly rainband data inferred that major spiral bands may grow at the expense of nearby convection, and subsidence of the outflow from their tops may keep the corridors relatively cloud-free.

It was during my mission on September 10, as the storm drew a bead on the Northeast, that we were joined by Edward R. Murrow and his CBS network *See It Now* film crew. Since Hurricane Carol had soaked New England the week before, interest in hurricanes had peaked, especially north of the Mason–Dixon line, and Murrow gave his audience a personal glimpse of what the Air Force hurricane hunters went through in fixing storms. Before the flight the film crew interviewed the flight crew and gave me an opportunity to explain to the television viewers why we should research hurricanes. It was during the broadcast of this episode that Murrow made his famous soliloquy: "In the eye of a hurricane, you learn things other than of a scientific nature. You feel the puniness of man and his works. If a true definition of humility is ever written, it might well be written in the eye of a hurricane."

This mission was my last opportunity in 1954 to fly a piggyback experiment, but the data obtained from these flights were not only invaluable in understanding hurricane structure, but the experience of drawing up the research operational plan (including the development of a cloverleaf flight track), developing a data acquisition design, and setting of objectives informed my latter efforts in organizing the National Hurricane Research Project. And these lengthy visits to Kindley Air Force Base in Bermuda helped me to develop camaraderie with the flight crews and the meteorologists, engendered in them an eagerness to acquire essential research data, and encouraged special flight efforts on their part.

While Edna was my last chance to fly a mission in 1954, it was hardly the end of that eventful hurricane season. In mid-October, Hurricane Hazel made a right turn in its westward track through the Caribbean and roared through the Windward Passage to make a big impact on the Carolina coast. It left a path of damage all the way to Canada. For the past decade, hurricanes had mostly affected Florida and the South; however, in 1954 each of the Hurricanes Carol, Edna, and Hazel affected New England and the mid-Atlantic states. All were important, notable hurricanes, and they inflicted damage affecting the constituents of more senators and congressmen than in any preceding year in memory. There was an open-mindedness, if not a sense of urgency, on the part of Congress to see that hurricane research was accelerated and well supported. In fact, the Weather Bureau, despite rejection of its requests from earlier years, was derided and made fun of for not having been more energetic in pushing hurricane research.

In late 1954, with Congress insisting on early action, we started reworking the proposals from earlier years, enlarged them in concept and funding requirements by a factor of 3 or 4, and presented Congress what we thought was a proper response to the urgency they were assigning research on hurricanes. It was proposed to set up a three-year project to begin with and then to reevaluate to determine the need for a permanent laboratory in terms of the progress made during those three years.

Chapter 6
The National Hurricane Research Project

Project Plans

The need and justification for mounting an all-out effort to better understand and design better scientific methods for predicting hurricanes was stimulated in the scientific community by the promising findings and outcome of the research flight of 1947 near the Bahamas, the one into Typhoon Marge near Guam, and the piggyback flights out of Bermuda in the early 1950s. Based on these studies, a plan for a three-year research program dedicated to aircraft probing of hurricanes and research to better understand, predict, and warn of hurricanes was developed and submitted with the Weather Bureau budget starting in 1952.

However, the request for research funds failed, mainly because of the lack of support within the administration's budget office and thus never reached Congress. It was a different story, though, in 1954 when three disastrous hurricanes (Carol, Edna, and Hazel) ravaged the New England and mid-Atlantic coastlines. These seriously affected the constituents of more senators during a three-month period than ever before. Led by Senator Theodore Green of Rhode Island, this congressional delegation pushed through a comprehensive bill, the largest outlay of federal funds ever for research in meteorology to that date, supporting plans for a National Hurricane Research Project (NHRP) based on an enlargement of the research plans developed several years earlier by the Weather Bureau.

The plans now called for a three-year effort to collect data on hurricanes from instrumented aircraft flying into the storms, from constant-altitude balloons released into the eye, and from suborbital rockets taking pictures of these storms from space. In addition, differing methods of analysis would be explored to find the best methods of presenting data to forecasters, and even possible weather modification strategies would be tested. This activity would be centered at a research operations base (ROB) at Morrison Field in West Palm Beach, Florida (later to become Palm Beach International Airport). The site was selected over collocation with the Miami hurricane forecast office (my preference) because we were going to be using Air Force aircraft, and it was more efficient and less expensive to operate from one of their airfields. Money was set aside to install the latest observational technology aboard three Air Force aircraft, and funds were also allocated for researchers in academic circles to exploit the information gathered by the project to improve theoretical understanding of tropical cyclones.

The Return of the Mauna Loa Observatory

Project funds ended up not only supporting the establishment of NHRP but also helped reestablish the Mauna Loa Observatory (MLO) in Hawaii. The circumstances that resulted in the reestablishment of MLO, however, read like a fairy tale. This effort hung on the chance encounter of two scientists at a site thousands of miles removed from Mauna Loa, two scientists who quickly discovered a common need. And in the course of a single exciting day of discussing and scheming, they agreed on a cooperative plan of action to be pursued by their respective agencies. The National Bureau of Standards and the Weather Bureau were persuaded to agree on a collaboration to reestablish the observatory, this time larger, better equipped, and continuously habitable with a serviceable access road.

In May 1955, I took the longest vacation I ever had while serving the Weather Bureau, five weeks, in which I took my two daughters to the canyon country in the western United States. Along the way we stopped at National Solar Observatory at Sacramento Peak, where astronomical observations were being made. There I met a scientist from the National Bureau of Standards named Ralph Stair, who was trying to make solar flare observations under very adverse conditions. During the weeks he had been there, he had only had three days of good observing weather. Stair and I soon found we had a number of common interests.

I suggested, "Why don't you go someplace where you can conduct observations more than 350 days a year?"

"Is there such a place?" he asked.

"Oh, yes," I replied. "At Mauna Loa summit. We had an observatory there for a while, which could be reopened with a little help."

"Tell me more," he said.

I explained, "The biggest problem is building a permanent access road and a building suitable for continuous occupancy at the site. The Weather Bureau is hamstrung in contracting for such tasks, which must be done through the U.S. General Services Administration (GSA) and usually turns out to be prohibitively expensive. However, I happen to know that the Bureau of Standards has enabling legislation that allows it more freedom to do such things. If we can work together, we may both be able to accomplish our objectives more easily than we could alone. If you, on behalf of the [Bureau of] Standards, would become prime contractor for the project, build the road and observatory building suitable for continuous occupancy, then purchase and install a Dobson ozone instrument, and measure total path ozone at the site for a minimum of two years, I think I can justify diverting funds from the NHRP to cover costs of the road and the building."

We both agreed that at first blush it looked feasible and that we would pursue it with our respective bosses. When I returned to Washington, D.C., I had a rougher time with Reichelderfer on this matter than I expected. He was concerned about "congressional intent" for use of NHRP funds, and he doubted the connection between ozone measurements and cold lows could be credibly related to tropical cyclone research.

At this juncture, however, unexpected support came from Harry Wexler, director of Meteorological Research, who said, "This sounds great. And we could measure carbon dioxide there, too. It is a site completely free of upstream pollution and at elevation that frees it from local pollutants. This truly could become a baseline station for observing and measuring increases of CO_2 in the atmosphere."

Work began very shortly, under the direction of the National Bureau of Standards with collaboration of Wexler and his staff. The new observatory on the slopes of Mauna Loa, 2000 feet below the summit site, was dedicated and placed in operation little more than two years after closing of the old summit site, another advancement of science owing substantially to the pursuit of cooperation.

From the very day of its dedication, the Mauna Loa Observatory was a creative factory of scientific achievements, including the initiation of its half-century records of daily weather variations, what became the world's longest record of atmospheric transmission as well as the variable constituents of the atmosphere at the elevation of the observatory. The new observatory shortly accumulated a bevy of "firsts," including discovery of Asian dust and other pollutants transported across the Pacific to North America. The activities and achievements of MLO, decade by decade, are a unique testimonial of what has been and can be accomplished by exploiting such facilities as the observatory affords for scientific explorations and research. It offers these not only to a dedicated resident staff but also for collaboration with talented scientists the world over, attracted to MLO's unique venue. It's a facility whose achievements and cost effectiveness are not only more than competitive with similar facilities with less suitable locations, but in my view it has only scratched the surface of opportunities that lie ahead, especially when the means can be found to reestablish the summit facility as a prime site for atmospheric observation and measurements, with a hard-surface access road with year-round sustainability connecting it with the slope facility. I remain as convinced now as I was in 1949 that the summit site, for many observations and measurements, is significantly superior to that at the slope site: air sampling at the summit should retain the character and substance of planetary flow, while that at the slope site tends to be sullied in content as a flow controlled by the mountain itself.

Herbert Riehl's Aspirations

The formation of NHRP was a feather in Reichelderfer's cap but also a challenge, not only to get the project underway promptly, but to do it right in the perception of his colleagues and supporters as president of the World Meteorological Organization (WMO). Reichelderfer's quandary was not only did the WMO require a research plan and objectives that met the approbation of its international leaders, but also the selection of a recognized tropical meteorologist as scientific head of NHRP. As best as I could tell, Reichelderfer was comfortable with the overall research concept but not in picking a distinguished leader.

Herbert Riehl of the University of Chicago was becoming widely recognized as the father of modern tropical meteorology by dint of his recently published textbook on that subject, *Tropical Meteorology*. When it was clear NHRP was to be funded, Herb immediately coveted the position of project director. From his publications and outstanding leadership in tropical meteorology, such an appointment was justified. But Reichelderfer, in evaluating the potential leadership capabilities needed in dealing with the concerns of both the scientists and the political entities the director would have to confront, promptly rejected the candidacy. Herb's pique at this result caused him to decide to have nothing to do with NHRP.

Reichelderfer knew that I knew that despite my involvement with hurricane research and the papers I had published on tropical meteorology, I had not earned a doctorate and therefore would lack credibility as scientific director of such a massive project as NHRP. However, he did give me the responsibility of presenting the current plans and objectives to many distinguished research scientists of the United

States, Europe, and Japan, and for recording and evaluating their reactions and suggestions. In the end (reluctantly, I surmised) he selected me to direct NHRP. With that decision made, plans for launching the project in 1956 proceeded rapidly.

The first order of business was the identification of the aircraft types that could safely achieve the kind of data acquisition efforts that would serve the project objectives. For this we turned to MIT to run computer checks of military aircraft that could fill the bill, and we presented that to the Navy and Air Force to propose the kind of support aircraft that would achieve our objectives. With the astute help of Professor Victor Starr at MIT, this first step was completed expeditiously. The Air Force, with "encouragement" from Congress and astute maneuvering from Air Weather Service's General Thomas Moorman, agreed to supply two B-50 aircraft and a B-47 jet to support NHRP.

Simultaneously, the bureau appointed the leading staff for the project and selected and equipped a project base at Morrison Field in West Palm Beach, while instrumentation of research aircraft was contracted out. The main management team consisted of Cecil Gentry (assistant director) from the Hurricane Warning Office in Miami; Noel LaSeur (associate director) from Florida State University in Tallahassee; Art Johnson (head of data processing) from Washington headquarters; Bob Rados (instrumentation specialist) from the Air Force Cambridge Research Laboratories (AFCRL) in Boston, Massachusetts; and Art Youmans (business manager) from Washington headquarters.

In May 1955, this group of project leaders together with an invited distinguished meteorologist, Joachim Kuettner, participated in a familiarization and feasibility flight to check out supporting facilities and flight procedures along the principal routes that we would expect to use on research flights, including Puerto Rico; Bermuda; Boston; Washington, D.C.; and Cape Canaveral, Florida, using a B-29 aircraft supplied by AFCRL and instrumented by Bob Rados.

The Research Operations Base

The research operations base of NHRP was opened in April of 1956, located in a warehouse on the north side of Morrison Field. After a settling-in period daily operations were started, with staff around the clock collecting, plotting, and analyzing weather information from across the tropical Atlantic. Daily map discussions were used to plan future operations, and a number of different maps were tested to see which gave the best indication of hurricane formation or future movement. In addition to the staff, there were frequent visitors from Florida State University, the University of Chicago, and other academic institutions as well as the Miami hurricane forecast office, which made for lively talks and interactions between researchers and forecasters. In retrospect, it seems remarkable that all these measures were completed by April of 1956 as a reaction to the ravages of three hurricanes as late as October 1954.

The Air Force promptly delivered the three aircraft (two B-50s and a B-47) for modification and new instrumentation. This included installation of probes and digital recorders, to be done at White Plains, New York, by General Precision Laboratories, a fortunate choice for the task. It was a credit to all concerned that these uniquely equipped research aircraft were ready for flight into the first hurricane of the season early in August 1956. This may sound like a straightforward, if not simple, task. But it was not! To meet such a schedule as we had set for ourselves, out of necessity, a hurried pace to get going while the iron was hot required not only dedication but ingenuity and a willingness to cut ruthlessly through normal leisurely bureaucratic procedures and roadblocks, not only in government, but among our contractors. But with the momentum generated by this time, the project leaders were well motivated to get the jobs done, and the few who were not were replaced (and there were several!).

Joanne Malkus at West Palm Beach

Once the National Hurricane Research Project had been funded in late 1954, and earlier plans based on results from the piggyback research missions had been adjusted to accommodate the broader objectives of NHRP, it was necessary to seek the advice of the acknowledged experts in tropical meteorology across the nation and abroad, a task which, by default, fell to me because a project director had not been named at that time. This is when I became reacquainted with Joanne Starr Malkus. In the years since I knew her at the University of Chicago, she divorced Victor Starr and married Willem Malkus, and they had two children, David and Steven. As chance would have it, Joanne, who was employed by Woods Hole Oceanographic Institution in Massachusetts, appeared in Washington, D.C., to represent that institution as its expert in tropical meteorology. Her able participation in this review process was the beginning of a remarkable and continuing professional and, soon after, personal relationship.

From the first hours we spent together reviewing plans and early results from NHRP, it was clear we had similar thinking in our concept of the hurricane, its puzzlements, and how best to seek a better understanding of both the dynamic and thermodynamic processes at work in creating and sustaining the destructive fury that distinguishes it from other storms. The starting point, however, differed considerably. Joanne's starting point was the fair-weather cumulus cloud, reaching out to the environmental setting that could make a deadly lion out of a pussycat. My starting point was the hurricane itself, reaching out for an understanding of how the structural members of the hurricane could have evolved from such a benign setting in the tropics. From these opposite starting points, we were asking essentially the same questions: Did the triggering process originate from local instabilities that fed on an environment of opportunity, or did the larger environment of opportunity depend on external dynamical impulses that concentrated the latent energies to form the hurricane? Whether we began at the top or bottom, we were on the same page.

While Air Force regulations did not permit Joanne, because of her gender, to participate in B-50 research flights, she spent a major part of her time at NHRP's West Palm Beach research operations base. The two of us could frequently be found, even after midnight, analyzing and discussing the latest research flight data, often preceded by dinner at the air base officer's club, and on Wednesday and Saturday evenings, a few turns on the dance floor.

With the help and encouragement of Joanne Malkus and the enthusiasm of the Finnish meteorologist Eric Palmen, Herb Riehl was persuaded to visit NHRP at West Palm Beach. After a couple of research flights I arranged for him to plan and direct, Riehl became an ardent supporter of NHRP and later welcomed me as sponsor of my doctoral research. Of course, Joanne working in collaboration with Riehl was able to delve much more deeply into the theoretical aspects of the problem than I could. My limited hours available for research were devoted to applying their findings for a better understanding of the prediction problem as well as to directing the efforts of NHRP to that end.

Joanne and Herb coauthored "Some Aspects of Hurricane Daisy, 1958," a landmark research paper on the energy transactions of Hurricane Daisy, examined by project scientists in 1958, and the processes that drove and sustained hurricanes in general, the first of a series of research efforts based on data acquired by NHRP research aircraft.

NHRP Operations Begin

The first research flight into a hurricane for NHRP occurred on August 13, 1956, as Hurricane Betsy moved over the Turks and Caicos Islands and threatened south Florida. I was the lead scientist on the inaugural mission and things went less than smoothly. Just an hour out, an alternator on the aircraft blew, taking out the radar, altimeter, and the automatic navigation system. Soon after it was discovered the wrong system was being used to calculate the D value, an important parameter for understanding what pressure level we were at. I decided to continue with the mission, even though we weren't recording any data of value. An Air Force radar station on Caicos directed us into the eye and we returned safely to West Palm after 2 p.m. only to find out that Gordon Dunn, who had been in charge of the National Hurricane Center in Miami just one year, had forecast landfall just south of the Palm Beach area.

However, as the hurricane reached the central Bahamas, it recurved and the mainland never got more than 25 knots of wind (1 knot = 0.51 meters per second). On the very next research flight, directed by Noel LaSeur, everything seemed to work except the recording of longitude and the IBM card count (we were recording digitally in terms of shaft position for each probe output, and recording values each second on IBM punch cards). When LaSeur's flight returned, there were grins from ear to ear as the crew clamored to the tarmac, flushed with success of the mission. But as he, the punch card operator, with four boxes of punch card records from the flight, stepped out of the plane, he missed the last step, dropped the boxes, and the cards flew everywhere under the wash of the propellers, which were still turning. Since the card count hadn't been working, it was no simple matter to put the data back in sequence. It took about three years of graduate student sleuthing to restore those cards to proper order and to obtain a printout of data suitable for research analysis.

We had three flights into Betsy. But the 1956 hurricane season was very quiet. In all, our first research season saw only 15 sorties into various tropical systems. After the devastating hurricane seasons of 1954 and 1955, 1956 had been a pussycat. As happens during nearly every federal administration, a year comes along when economic necessity places every agency's budget in jeopardy, and it becomes a game of cat and mouse as to which line items are to suffer. Since nothing earthshaking was accomplished the first year of NHRP, a number of people not only in the Weather Bureau but also at the Department of Commerce and the Bureau of the Budget (now the Office of Management and Budget) looked longingly at the large amount of resources devoted to hurricane research as an attractive target for economies, and for it to serve as a cushion for longer-term programs that otherwise would suffer. Rumors were rampant that NHRP was on the chopping block to lose the biggest part of its research funding. Dr. Reichelderfer was concerned about it, as were others in the central office, who saw to it I was kept advised. Others in the central office felt entirely too much money was being spent on hurricane research while the rest of the Research Division was losing ground. So, we had both internal and external pressures acting to cut NHRP off at its roots.

Something had to be done or we were going to lose the project, or have it emasculated to the point of ineffectiveness. With this in mind, more than a review of NHRP progress during its first year, I decided to call on short notice the first Technical Conference on Hurricanes, not an American Meteorological Society–sponsored meeting, but one sponsored by the Weather Bureau at Palm Beach. First, we invited the lay users, people who would benefit most from the research results, such as the vice president for safety at Dow Chemical in Freeport, Texas, and key figures from offshore oil drilling programs in Texas and Louisiana. Second, we invited those from universities and scientific agencies interested and concerned about NHRP's plans and programs.

During the three days of meetings at West Palm Beach, I was taken aside by the vice president of Dow Chemical, who commented, "I am absolutely delighted to see NHRP got off the ground. However, I know along the way you are going to run into political problems of one kind or another. If I can ever help out, let me know."

I replied, "As a matter of fact, we are having problems right now. Our problems, however, are with the White House not with the Congress. The Congress, by and large, I think, will back us. But there is no way that I know of to get to the White House."

He said, "Well, I do. I happen to be the president of the lobbying group that includes not only Dow Chemical but also Westinghouse, Monsanto, and General Electric, and a few other large corporations. We don't deal with Congress; we deal with the White House. We are meeting two weeks from today in Washington, and I am going to take this thing up and see what can be done." He made good on his promise, and the Bureau of the Budget made a 180° turn; we got continued support, somewhat larger than actually expected.

Dr. Reichelderfer never found out what I'd done. If he had, he probably would have had my scalp, because that kind of political activity was a no-no for employees not in policy positions. But I took the risk and decided that it was worth putting my neck on the line for a program that was doomed without such action. Fortunately, luck has favored me throughout most of my career, and stood by me in this instance from which I believe all have benefited.

The following year, Hurricane Audrey was the most memorable event, memorable because more than 400 lives were lost—lost not because of prediction failure but because of the failure to communicate properly with coastal residents. The forecast, it turns out, was quite good and timely, as were the issuance of warnings. Unfortunately, the warnings and advisories issued by the Weather Bureau, normally distributed by radio broadcast from a station in Lake Charles, Louisiana, were edited by newly elected city officials, who eliminated all but what they considered of importance to Lake Charles, forgetting that coastal residents needed the full text of the advisory to make their evacuation decisions. The result was that most people residing on the coast gained the impression they need not evacuate until morning. When they awoke in the middle of the night with floodwaters rising, all evacuation routes had been cut off.

While this led to a long and protracted lawsuit against the Weather Bureau (finally resolved in favor of the bureau), it did not influence adversely the NHRP. In fact, it gave it visibility, which was quite beneficial. Because of the importance of Audrey, I decided on short notice that the project should be represented on the scene as quickly as possible. So, we dispatched our logistics support plane with a small group of scientists, who reached Lake Charles about 30 hours after landfall. Once there, I persuaded a military helicopter unit to transport us over the coastal region to survey high-water marks and damage and to talk to residents about their reaction to warnings. This not only produced much useful information of scientific value but also of value in reviewing and revising the structure and semantics of advisories and warnings.

Moreover, it proved valuable in obtaining early estimates of the scope and maximum heights of the storm surge. Lee Harris, who had been appointed NHRP project leader for storm surge investigations and predictions, participated in the survey of Audrey and was especially impressed with the value of early surveys in studying storm surges. Harris and his successor, Chester Jelesnianski, went on to develop the excellent Sea, Lake, and Overland Surges from Hurricanes (SLOSH) model for predicting storm surge expectancies, a model still in operational use today.

It turned out to be the first of many such helicopter survey trips to the disaster scene made by NHRP and later by the National Hurricane Center. In fact, I participated in surveys of every major hurricane that crossed the U.S. coast from 1957 to 1972, perhaps the most important of all being that of Camille in 1969, where we were able to identify the place of maximum surge and to calculate a record height of 26 feet, only one foot off the final surveyed value of 25 feet.

1958 was the third—and supposedly final—year of the NHRP. The Air Force had to be persuaded to continue cooperation with the use of their aircraft and personnel for this season and it was becoming obvious that this would be their last year. But it was obvious to all the scientists involved with NHRP that we were only beginning to scratch the surface of understanding the structure and dynamics of hurricanes. Luckily, the 1958 hurricane season offered us more research opportunities than the last two combined, and we flew 41 sorties into all sorts of tropical cyclones, most importantly, Daisy, which Herb Riehl and Joanne used in their energy budget study.

NHRP's future

When the Air Force advised it would not be able to support NHRP with aircraft after the 1958 season, the question was whether there was justification scientifically for acquiring other research aircraft and support facilities. Why not stop and just do research on the data we obtained with the Air Force planes over the first three years? After all, the sense of urgency for hurricane research generated by the 1954 and 1955 hurricanes had waned considerably, and to procure and instrument new aircraft or reequip old ones would involve significant additional appropriations at a time when new funds were hard to come by.

It was perhaps a greater and more difficult challenge by far than the original one in persuading the Air Force to support NHRP. This time we faced not only the external political problem of getting budgetary support but also significant disaffection with the project within the scientific community. A polarization of opinions had developed between those who perceived that the project as a sink for research funds without evidence in the literature at that juncture of useful results from the mountains of data acquired and those enthusiastic supporters who had participated in one way or another in the data acquisition or had been gratified with the data as they applied to their research. Fortunately, the supporters were more articulate at the right times and places than the detractors but this advantage wasn't enough. We had to demonstrate, through a credible review, that progress was made but vital questions were left unanswered. If the expenditures to date were to pay off, we simply had to continue taking advantage of new technology that had become available in the interim to continue data acquisitions. One of the questions that had reared its head was in regard to the computations of energy budgets based on winds measured by the aircraft's Doppler systems, which suffered during turns of the airplane, resulting in "a moving ocean platform" from which signals were reflected.

To bolster my arguments in making a case for continuance, I enlisted the strong voice and experience of Verner Suomi in evaluating the performance of earlier instrumentation and recommending in a summary statement the new technology that should be used on the replacement aircraft. I also returned to MIT and got a reevaluation of aircraft most suited for our purposes. Finally, I got a contract with General Precision Laboratories, the company that installed our probe and recorder systems on the Air Force planes, to generate specifications for the new instrumentation system recommended by Vern to provide the basis for bid invitations. With this in hand, I pulled out all the stops in launching the campaign for continuing the data acquisition program of NHRP. This included recommendations for the purchase of three new C-130 aircraft at $4.5 million apiece (in those days), equipped with inertial navigation systems, a high-speed solid-state Ampex recorder system, and new meteorological and cloud physics probe systems using the very latest technology.

The campaign, of course, began at the central office, where opinion was somewhat divided. However, I got the nod to proceed in principle to determine the support from the scientific community. This was done by staging an American Meteorological Society (AMS) technical conference at the Deauville Hotel in Miami Beach to which we invited a good mix of scientific expertise from across the nation

Following the presentation of papers, there was a formal review hearing regarding the project's usefulness. The upshot of the review was a resolution, signed by such luminaries as Eric Palmen, Jule Charney, Sverre Petersen, and Herb Riehl, asking the Weather Bureau to continue support for a data acquisition program of hurricanes. It generated sufficient support to persuade the Weather Bureau to submit a supplemental appropriation request through the Department of Commerce that was far larger than any previous one from the bureau. That conference, by the way, became the first AMS Hurricane and Tropical Meteorology conference, now held every other year.

I never really expected to get the three C-130s, but I felt that when this request was rejected, not by the Department of Commerce but by the Bureau of the Budget, that the alternatives would reduce the budget so much that the remainder of the package would make it through unscathed, which is essentially what happened. We were fortunate to find an airline company, known then as Trans Caribbean Airways, that had just bought two brand-new DC-6As that they had to sell quickly as a result of an opportunity to obtain new jet aircraft they originally thought they couldn't afford. We purchased the DC-6s for less than $500,000 apiece. The Air Force then came through again, agreeing to a bailment to the Weather Bureau of a B-57 Canberra for high-level flights and a small logistics support aircraft.

All that remained was to solicit bids for the modification of the aircraft. We held a bid invitation conference at Palm Beach in which the specs were presented and all the options explained. With all this behind me, I shoved off for Chicago to complete my doctoral degree, feeling assured that General Precision Laboratories would be awarded the contract and all would be well.

Unfortunately, that wasn't the way it worked out, although I was necessarily out of the loop for the follow-up and unaware of what was happening. Of the six contractors whose bids were validated, General Precision's bid was competitive with all but one, a Chicago firm with large Department of Defense contracts whose bid was less than half that of the next lowest bidder, an omen in itself that something hokey was going on. And here administration politics struck a blow for mediocrity.

The legal staff at the Department of Commerce, reviewing the bids and the Weather Bureau's recommendation not to accept the lowest bidder, not only required that the lowest bid be accepted but, without consulting the bureau, wrote the contract to provide a cost-plus-fixed fee payment schedule.

After six months at work, the Chicago company had spent two-thirds of our budget for instrumentation and wasn't nearly half finished. About that time I received a call in Chicago from Russ Grubb, the head of the Weather Bureau's budget office, asking if I had any suggestions regarding the dilemma. After further discussions with Reichelderfer, I was authorized to assemble an independent team to investigate the problem and recommend action. I then persuaded Horace Byers and Roscoe Braham of the University of Chicago and Vern Suomi of the University of Wisconsin, all three veterans of successful field programs, to join me in the investigation. The result was a recommendation that the Chicago company be defaulted and a new sole source contract be negotiated with ESS GEE company of Tarrytown, New York, to complete the job. This latter firm was staffed by the General Precision engineers who had developed the specs originally and had subsequently split off to form their own company. The recommendation was accepted, and Vern Suomi agreed that he and a colleague, Tom Parsons, would monitor the contract. Fortunately, the overhead of ESS GEE was only a fourth that of the Chicago firm, so that the job was finished and accepted essentially within the original budget. Incidentally, as an outcome of the default, the Chicago firm was barred from defense contracts for three years, and the company's president and two other senior officers were fired.

I was then able to return to my studies, leaving NHRP in the capable hands of my friend Cecil Gentry. The project was able to move south to Miami to be collocated with the Miami hurricane forecast office in their new lodgings at the Miami Aviation Building, near Miami International Airport, where the DC-6s were housed. This close grouping and collaboration of operational forecasters, hurricane researchers, and aircraft experts was what I had originally conceived when I drew up the NHRP plans, and it came to be known as the National Hurricane Center.

Chapter 7
My Doctorate and STORMFURY

Back to Chicago to Complete a Doctorate

After the 1958 field program at NHRP, I felt that the project, its objectives, and its procedures were sufficiently well established that little would be lost by turning the day-to-day management over to the very capable assistant director, Cecil Gentry. There was no question where I wanted to complete my education. I wanted to work under Herb Riehl, who had invited me to return to the University of Chicago and let him supervise my doctoral work. It's possible, though I doubt it, that I might have felt a little different about returning to Chicago had my alma mater, Southwestern University, decided a little earlier to award me an honorary doctor of science, which they did in 1963 after I had completed my doctorate.

From my experiences at NHRP, I convinced myself—and later Dr. Reichelderfer—that effective collaboration between university research groups and the government researchers made it necessary that the Weather Bureau mount a program of completing the education of key research and operations leaders. I asked to be the first to be sponsored under the new program and Reichelderfer agreed. Later, when I returned to Washington, D.C., as Harry Wexler's deputy, I was project leader in implementing this program, and the first two people who returned to the university for this purpose were Cecil Gentry and Harry Hawkins of NHRP.

I happily arrived in Chicago the first of January 1959. I stayed there through June 1960, completing the course work and getting my prospectus accepted. I had been given two years to complete my degree, and that limitation greatly restricted my choice of dissertation subjects. Herb Riehl, my doctoral sponsor, had suggested I consider investigating the impact of ventilation of the hurricane core by entrainment of environmental air as a subject, and together we published an introductory paper that gained very little attention at the time. While I found the proposed subject very appealing, in writing the introductory paper (and later the outline) of a prospectus for this investigation, it appeared there was no way I could complete an acceptable dissertation in the time given me to do so. It appeared to be a Pandora's box, requiring endless follow-on research efforts that likely would be demanded by the sponsoring committee. Riehl agreed. So I chose instead an easy subject for which a database already existed.

Early in 1960, the U.S. Navy ran Operation Skyhook in the Caribbean Sea, launching high-altitude balloons bearing cosmic ray detectors to measure this mysterious radiation above the thicker layers of the lower atmosphere. In support of this experiment, the U.S. Weather Bureau's West Indian radiosonde network launched its weather sounding instruments with special plastic balloons, allowing them to rise far higher than the regular rubber balloons. Their readings gave an unprecedented look at the middle part of the atmosphere, the mesosphere (from about 30 to 60 miles high), which hadn't been regularly measured by radiosondes, aircraft, or instrumented rockets. And Herb Riehl had access to this unique dataset.

My doctoral dissertation examined the structure and motion of a wind shear line that moved across the experiment area from mid-January to early February of 1960. "On the Dynamics of Disturbed Circulation in the Lower Mesosphere" was accepted in 1962, and I finally had my coveted doctoral degree. It was published as one of the National Hurricane Research Project's reports in August of that year.

Meanwhile, Joanne and her husband, Willem Malkus, had accepted professorships at the University of California, Los Angeles, moving with their three children—David, Karen, and Steven—to Los Angeles. Shortly afterward, she took a temporary assignment as professor of meteorology at the University of Chicago, which quite fortuitously coincided with my deployment there. That year, 1959, was a delightful resumption of our earlier associations in Florida, despite the academic pressures I faced completing my degree.

Return to Headquarters in D.C.

When I left Chicago in 1960, rather than being returned to NHRP, I was assigned back to Washington, D.C., as deputy directory for severe storms of what was then the Research Division, headed by Harry Wexler. The writing of the dissertation, which was to be done on the job at the Weather Bureau, was completed in time for the university's August convocation in 1962. My two and a half years as deputy director were filled with new challenges and 16-hour workdays. During this time NHRP acquired its own fleet of research aircraft; it moved to Miami and was adjacent to the Miami hurricane forecast office and designated the National Hurricane Center.

Both were relocated to the campus of the University of Miami the following year. In my earlier years, working as an assistant to Reichelderfer, my travels to forecast centers across the country convinced me that the Weather Bureau should promote a policy of relocating forecast centers to college campuses every time an opportunity presented itself, not because of the academic atmosphere, but as a means of establishing a mutual awareness of mission-oriented needs for new knowledge and for early incorporation of research results into operations. The advantages of such a relationship had been well documented at the University of Chicago during the years Gordon Dunn was in charge of our forecast office on that campus.

Also the local severe storms research unit in the Weather Bureau's office in Kansas City, Missouri, was moved to the campus of the University of Oklahoma in Norman, where it was reestablished as the National Severe Storms Project (NSSP) under new director Ed Kessler. One other goal of mine always was to keep the research laboratories in close adjacency with related operational units. We succeeded in doing this in 1958 when NHRP was moved to the Miami hurricane forecast center. Moving NSSP away from Kansas City seemed to be a rejection of that concept, but it was only a compromise of the moment that was required, first, for logistical reasons and, second, because of a personnel management problem. The compromise was mitigated by moving NSSP into proximity of, and into close working relations with, the University of Oklahoma. Some have questioned whether NSSP was as effective in this regard, but that was the goal we set, and I give Ed Kessler credit for his effort toward this end. Eventually, the Severe Storm Forecast Center, renamed the Storm Prediction Center, was moved to Norman in 1997.

I also arranged for NHRP's research aircraft to be shared with NSSP when not required for hurricane missions. Eventually both projects were upgraded to laboratory status, ensuring that they would be permanent efforts of the U.S. government to improve our understanding of dangerous weather. Their names and acronyms would thus change to the National Hurricane Research Laboratory (NHRL) and the National Severe Storms Laboratory (NSSL).

Riehl's Unintended Contribution to the STORMFURY Concept

A surprising and unintended result of Riehl's activities at NHRP ended up planting the seeds that led to the establishment of Project STORMFURY, an experiment to explore and determine the feasibility of reducing the destructive potential of a hurricane by weather modification. I had returned to Miami in the summer of 1960 to participate in the hurricane field program. The leased DC-6 aircraft finally had been instrumented and become available to the project, providing plenty of opportunities to collect data. In addition, Herb had managed to secure a seat on a high-level flight aboard a Navy jet above the strong Hurricane Donna in September 1960.

In describing his experiences to me later, he stated that the sole source of hurricane outflow clouds was confined to the right-front quadrant of the eyewall, spreading downstream several hundred miles. Suddenly a light came on, shining through the fog of my memory and illuminated various other—mostly failed—efforts to modify the weather by cloud seeding methods. Well aware of the anathema with which Riehl regarded weather modification, my responses and queries about his Donna descriptions were rather guarded and the subject was soon changed. After all, I was in the midst of preparing my doctoral dissertation on quite an unrelated topic and Riehl was my sponsor!

But the light continued to elucidate the subject, NOT primarily for the humanitarian outcome it posed, but because of the (not always well disciplined) discussions and arguments I had engaged in (and had rarely prevailed at) during the weekly seminars at Weather Bureau headquarters during the 1940s with the likes of Harry Wexler and Jerry Namias concerning the development and steering of hurricanes. The bone of contention was whether changes in development and movement of hurricanes were dependent solely on the momentum of the larger-scale environment or whether they were critically responsive to smaller thermodynamic irregularities on the cumulus scale. The latter concept was the one I had embraced all the way back to my New Orleans days when trying to understand why our predictions went awry and also from discussions later with Grady Norton. But years later, after my and Joanne's postmidnight sessions during the early days of NHRP, the realization that her baseline thinking about hurricane behavior, starting with the simple cumulus cloud, and mine, starting with the fully blown hurricane, conformed remarkably in many ways. So, off I went to get Joanne's take on what Riehl had seen, merging our concepts with new ideas.

The birth of STORMFURY

The National Science Foundation, through Earl Droessler, agreed to support the hurricane seeding program, which was aimed at reducing hurricane strength, and provided a grant to the Weather Bureau, based on the plans I had developed and included the crude hypothesis as basis for the exploratory effort. In the spring of

1961, as I was drawing up the plans for the modification effort, I was visited in my office by Pierre St. Amand, a talented chemist stationed at the Naval Ordinance Test Station (NOTS) in China Lake, California. He offered support by supplying newly designed cloud seeding pyrotechnic generators, huge canisters he had christened "Cyclops." Each were designed to fall in a slow windmill fashion, dispersing silver iodide smoke over a considerable volume surrounding its fall path. He also offered to obtain Navy planes to deliver and release the generators at the top of the hurricane, together with other planes for monitoring the results. It was almost too good to be true, but after careful inquiry in Navy circles, it turned out that he had the clout and the ability to deliver what he promised.

An informal, tentative agreement was reached between his office and mine for this collaboration. I should mention at this point that before the Navy entered the picture, our first difficulty with St. Amand occurred when we learned that he had his own idea of where and how to seed, and that his objectives were to steer, not influence the strength of, the hurricane. The controversy that followed, with Joanne Malkus supporting my position, reached the highest levels of the Navy. The Naval Weather Service sided with the Weather Bureau, so St. Amand was relegated to the task of supplying the pyrotechnics.

We were able to organize an initial test that summer. In September, Hurricane Esther moved into range of the island of Puerto Rico and a fleet of Navy and Weather Bureau aircraft were dispatched to Roosevelt Roads Naval Station in Ceiba, Puerto Rico. The first day was marred by the late takeoff of the seeding plane, which led to the canisters falling outside the design seeding area, one of them passing perilously close to one of the Weather Bureau's DC-6s. However, the next day the mission into Esther went well, in that structural changes in Esther following the seeding were recorded by the monitoring aircraft in accord with the experiment design, including the expansion of the eyewall diameter and the rising minimum pressures in the eye. These results were promptly reexamined and validated jointly by Joanne and myself.

Despite this uncertain start, we were encouraged by the level of cooperation in accomplishing a very complicated flight plan in such a short time. The Navy and the Weather Bureau soon reached an agreement to work together on hurricane modification, establishing Project STORMFURY in 1962.

This new project would be separate from NHRP, but it would be supported by its staff and would be administered from my office in Washington, D.C. (as deputy director of the Weather Bureau's Research Division). It would largely be funded by the National Science Foundation with aircraft support by the Navy and NHRP. I was project director and was assigned Cecil Gentry as deputy director, my friend Capt. Max Eaton was the assistant director, representing the Navy, and Joanne Malkus was my science advisor.

It was a rough but exciting ride getting STORMFURY underway, stoked by a curious mixture of gratification and frustration. Through it all, it not only brought Joanne and me closer together professionally and scientifically but personally and emotionally. We were two very different people, with differing family and personal backgrounds, differing ambitions and drives, but we rapidly bonded in a common urgency to support each other's goals in life effectively. And we did for nearly half a century during her life, which supported our togetherness.

Other severe storm projects

There were other considerations that also commanded my attention while getting STORMFURY off the ground. Over the previous decade, the military had been developing drones, remotely controlled unmanned flying vehicles. The Navy used theirs for targeting, but I saw the possibilities of using them as a research platform. They could gather atmospheric data from where it was too dangerous to fly manned aircraft. In collaboration with the Naval Research Laboratory in Washington, D.C., I participated with Robert Ruskin in the experimental launch and retrieval of minimally instrumented research drones into squall lines from a military base in the Oklahoma Panhandle.

Tornado and severe thunderstorm research at this time was being carried out on an ad hoc basis by Clayton Van Thullenar out of the Weather Bureau's Kansas City office, but the emphasis was mostly on operational applications. I was determined to change this. As was the case with NHRP, my first concern was that the research be done with a specific mission orientation, not applied research per se, but as basic research whose mission is to support over the long haul our service objectives. Second, that it be done in a manner that fostered partnerships with and encouraged participation from the scientific community at large, especially university scientists, including the awarding of grants for this purpose. I felt that NHRP was a good example of this.

I began making plans to start a national severe storms project and collocate it with the University of Oklahoma in Norman. But Mr. Van Thullenar didn't want to leave Missouri and in the end he was replaced by Dr. Edwin Kessler, who implemented the move and began a close working relationship with the university professors. The unit left in Kansas City became the National Severe Storm Prediction Center and emphasized using NSSP research to improve tornado and thunderstorm forecasting.

Another STORMFURY experiment

Our next opportunity to seed a hurricane came in August of 1963. Joanne Malkus was already at Roosevelt Roads Naval Station supervising a cloud seeding mission when Hurricane Beulah entered the STORMFURY zone northeast of Puerto Rico. The experiment began on August 23 with Joanne and I in one of the NHRP DC-6s. This was her first flight into a hurricane. During her previous participation with NHRP, the Air Force didn't allow a woman aboard one of its planes during what it considered hazardous duties. There were no such restrictions on the Weather Bureau research aircraft.

Operations ran more smoothly this time, even though we had over 10 aircraft operating in and around the hurricane at various times, including a high-altitude U-2 snapping pictures of Beulah's eye from 60,000 feet. Luckily the hurricane was moving slowly, so we had another day to try a second time. We were delighted to observe a lowering and fracturing of eyewall cloudiness on the radar screen as well as increases in the diameter of the eye, although as I told a news reporter, "it wasn't a knockout blow."

Management Changes at the Weather Bureau

It was during this period that Francis Reichelderfer retired as chief of the Weather Bureau and was replaced by Robert White, who had headed up the weather research wing of Traveler's Insurance. In the reorganization that followed in 1964, White was named director of a new agency known as the Environmental Science Services Administration [ESSA, which later became the National Oceanic and Atmospheric Administration (NOAA)]. ESSA administered not only the Weather Bureau but also several other federal science agencies, including the Coast and Geodetic Survey. George Cressman was appointed Weather Bureau director, and White created a separate research wing for ESSA that contained NHRP and NSSP. Both were made full-time laboratories and resulted in changing their acronyms to NHRL and NSSL.

Years later I was told confidentially, but perhaps lightheartedly, by a senior official that the top staff agreed that Reichelderfer, before his retirement, had been grooming me to succeed him as chief prior to Bob White's appointment. This came as a complete surprise to me, but it was something I could never have welcomed or accepted if offered. I was never cut out to serve with contentment the stultifying life of most top-level bureaucrats of my early acquaintance at 24th and M Streets NW in Washington, D.C., or of those who had served a decade or more in a single federal position, as I had once judged (unfairly) many of the old-timers in the central office. But whether or not the confidential information was valid, I never shared it with anyone, not even Joanne, and am grateful things turned out the way they did.

An Enduring Marriage to Joanne

With Bob White's asscension, I was named director of the Office of Operations. Overall, it was a less challenging and rewarding job than my position in Research at ESSA, with little perceived opportunity to make a difference. But it was a necessary transfer. Dr. White had offered Joanne a position within ESSA's research wing, which eventually led her to become the director of the Experimental Research Laboratory in Washington, D.C. This presented a problem because in January 1965 she and I had married. If I had remained as deputy director of Research at ESSA, then I would have been in a supervisory position over her and that would be a violation of nepotism statutes. As necessary as the move was, it was indeed a surprise and a disappointment that her new job cost me my research position.

But not to worry! As bureaucracies are often wont to do, a way around was found. Joanne was appointed STORMFURY's second director, since I was no longer in research. However, as initial director of STORMFURY, I was authorized to work and travel extensively in furtherance of the project's plans and objectives. Joanne and I cooperated in joint authorship of papers on STORMFURY and its progress.

Joanne and I were married (at the earliest moment we were legally entitled to do so) on January 6, 1965, at the county courthouse in Arlington, Virginia, with minimal outward fanfare except the joy in our hearts with the release of the inevitable tension that developed during our long wait for the news that we were at last eligible. The real celebration came following our church wedding a week later conducted by the minister of our Coral Gables Methodist Church in Miami, Dr. Hanger. He was one of the five or six ministers during my lifetime of attendance at the 17 Methodist churches where the sermons grasped my attention away from the silent worshipful queries and reasoning that usually claimed my attention.

Sabrina I

The second week of our marriage, Joanne and I visited the bank that held our meager joint account and arranged a loan of about $9,000 with which we purchased our first jointly owned sailboat, a 27-foot Bristol sloop, which was christened *Sabrina I*. It was launched and berthed at a small marina in front of our small apartment in Annapolis, Maryland. Its maiden voyage was made the following weekend. Off to a proper and uneventful start! We regularly commuted from downtown D.C. and sailed the Chesapeake Bay every weekend.

Joanne's STORMFURY Tenure

Joanne's tenure as STORMFURY director had a rough start. We coauthored several articles about the project, but one of them touched a nerve with Louis Battan of The University of Arizona. He thought our claims of success in the first two experiments were stating that we had positively altered the storm. All we were contending, however, was that the aircraft and personnel had carried out their assigned tasks according to plan. It took some time to clarify our point and smooth over ruffled feathers.

Then in late August 1965, Hurricane Betsy formed in the western Atlantic and was forecast to move into the predetermined area north of Puerto Rico where hurricane seeding was to be allowed. So, Joanne ordered the fleet of Navy and Weather Bureau aircraft committed to the project to deploy to Roosevelt Roads Naval Station to stand ready to carry out a modification experiment on the looming storm. On the day of the trial, however, Betsy stalled just south of the experiment area boundary. The early monitoring planes, with Joanne aboard, took off at their appointed time, but on their way to the hurricane, word came down from our boss, Bob White, that the seeding was cancelled because Betsy was still some 60 miles from the allowed zone.

Joanne had a hard decision to make about what to do next. Instead of calling off the operation, she decided to go ahead with the experiment as planned except no silver iodide would be released into the storm. This turned the effort into what is known as a "null" case, where everything is exactly the same as a real test except for the seeding. This could then be used to show what a nonseeded storm would do for comparison purposes. The planes executed their patterns and observed the unmolested Betsy before returning to base.

The only trouble was that the press corps back in Miami wasn't properly informed that the experiment had been called off. Later, when Betsy made some odd loops off the Florida coast, the newspapers ran stories blaming this behavior on hurricane seeding. It took a lot of explaining to correct the record, and there are people to this day who think Betsy was seeded and that the seeding had caused its eccentric track.

There continued to be criticism of the project among scientists as well as the popular press. There was also pressure from higher-ups in the government to hurry the experiments along so that practical weather control could begin. Joanne wasn't happy with all the negatives involved with being Project STORMFURY director, so she resigned the position in December of 1966 and went back to directing the Experimental Meteorology Laboratory, where she could carry out seeding experiments on isolated cumulus clouds without all the public scrutiny.

Reorganization of Operations

My time in Operations was not wasted, however. A decade before, I had tried to provide a research scientist at each forecast office to keep the forecasters updated on the cutting edge of new science and to help develop methodology for improving forecasts. It simply had not worked, primarily because these scientists found themselves isolated with little or no funds to maintain contact with universities or the scientific community. Second, it was because of the constant pressures to reduce expenses in operational areas. Regional offices, running short on funds, would insist that vacant forecaster jobs be filled with their research scientist, thus vitiating the entire program.

So, when I became deputy director of the Weather Service, Dr. Cressman and I agreed a better way would be to establish a scientific services branch at each regional office. This worked better in some regions than others, but it did not succeed in many instances in cultivating partnerships or active collaboration with university scientists, something I have always thought essential for the robustness of scientific research and its impact on operational missions.

Chapter 8
The National Hurricane Center

Gordon Dunn's Retirement

In 1966, Gordon Dunn announced his intention to retire as the director of the National Hurricane Center (NHC). With my weariness of the Washington, D.C., bureaucracy, much of it reminiscent of Weather Bureau headquarters in 1945, and with virtually no remaining opportunity for research in tropical meteorology, I asked to be reassigned to Miami as Gordon's successor. It was a request my bureaucratic colleagues found both strange and unpopular. After all, it was a considerable step backward in the civil service grading system, would disrupt years of prepping me for higher-level positions of responsibility and, as I was admonished by my boss, George Cressman (director of what had become the National Weather Service), it just isn't done in Washington. That came from a man I had strongly recommended for appointment as director. George, of course, had intended the comment as a compliment, but it instantly recalled my first visit to headquarters in 1945, and the conclusion it bred then was that this is what's wrong with Washington. But I went over Cressman's head to Bob White with my request. Bob White, the recently appointed administrator of the agency conglomerate soon to become known as NOAA, not only agreed to my transfer, but also my suggestion that Joanne's Experimental Meteorological Laboratory (EML) be relocated to Miami and be adjacent to NHRL and other NOAA research in that area.

Thus ended my exposure to the Washington torpor and the beginning of perhaps the happiest and most fulfilling period of my career of explorations in meteorology, nearly all of which was shared with Joanne. This togetherness, a unique mixture of professional, personal, and emotional experiences (some might suggest spiritual without arguing the definition of the latter), was one of unsurpassed happiness and exhilaration, and one of personal growth and maturity. It included notable successes but also a sense of personal fulfillment. It included fascinating and instructive worldwide travel to all this globe's continents, oceans, both polar areas, including a wide variety of the world's climates and cultures, safaris, and explorations in many remote areas, including mountain climbing in the Himalayas and blue water sailing. I am personally confident that it made me a better scientist and a better leader, and generally a better user of such talents as I inherited, than otherwise would have occurred. And I am arrogant enough to believe that these benefits applied in large measure to Joanne as well, though often in diverse ways.

When I went to Miami it was my primary objective—and in this I believe I succeeded—to bring to bear in hurricane forecasting what had been learned through NHRP of hurricane structure and energetics. This enabled the forecasters to diagnose and on occasion to challenge the validity of computer model predictions of hurricane movement, and to make intelligent, scientifically informed choices of various prediction options.

Separation of Research and Operations

NHC, before my arrival, engaged in more than hurricane forecast and warning services. It also covered aviation forecasts, daily weather, agricultural, and marine forecasts and advisories. A redesign of operational space accommodated a new satellite communications and advisory service. However, when NOAA's reorganized research arm was relocated to Boulder, Colorado, NHRL was directed to confine its research effort on basic research in tropical meteorology and curtail its applied research programs in support of hurricane forecasting, a first step in the separation of the togetherness and camaraderie that had unified these efforts between NHRL and NHC, and daily communication between Operations and related activities supporting the hurricane services program. This break was exacerbated when the research activity was relocated to a separate building miles away in 1984, one primarily serving NOAA's marine research program, responding primarily to a different drumbeat half a continent away with a competitive stance rather than with cooperation with common objectives.

Following the dispute over this action, NHC was authorized by the Weather Bureau to establish a hurricane specialist unit and a small research group to devote full time to applied research in support of the center.

The isolation of NHC from NHRL, even today in my view, never managed to serve any of NOAA's research or operational elements well, although the hurricane specialist unit provided program and policy management flexibility. The juggernaut of bureaucracy imposed from a distance was no match for the togetherness shared before Research and Operations were separated.

Return to Miami as NHC Director

However, as too often turns out to be the case, the career opportunities I enjoyed in returning to Miami ended up depriving a good friend of a lifetime goal he had sought. Cecil Gentry, my assistant director at NHRP and long-time personal friend and colleague, had eagerly sought to head up the National Hurricane Center, succeeding his mentors Grady Norton and Gordon Dunn. Indeed, he was considered the front-runner until I pressed my unexpected—and unwelcome—interests in the position. Ironically, it would have been my responsibility to appoint the successor to Gordon Dunn, and it properly fell to my lot to be the first to advise Cecil of what had happened. I called Cecil with the harsh news, without apology but with an appreciation of what it would mean to him. But it hurt both Cecil and me. Nevertheless, Cecil was gracious in accepting the outcome (though I came to believe his family was less forgiving.)

It was the second time in two decades I had had to confront such a self-analysis of motivation and justification of an action that would have a hurtful outcome either for myself or someone close to me, the choice of which resting solely with ME. Whatever the decision, the ultimate outcome, judged dispassionately, would probably be attributed to arrogance or selfishness, good analysis or bad judgment, with a high degree of uncertainty. In both cases, right or not, I passed the buck, seeking divine guidance, and the best objective analysis I could muster, ending up with what some may have judged a selfish solution. But in both cases, I ended up feeling comfortable that a greater difference resulted from the choices made than with the other options.

With our final NOAA transfer from Washington, D.C., to Miami and the university campus in Coral Gables settled, Joanne and I attended the annual hurricane review conference in Miami during January of 1967, and while there we purchased a lot for our new residence, selected a house plan, and contracted its construction for completion by the time we planned to arrive in May. Believe it or not (and we doubted it at the time), the house was ready for occupancy the day after our arrival. David and Steven (Joanne's sons) drove Joanne's red Volkswagen, while Joanne, Karen (Joanne's daughter), and I led the way in our Dodge family car on the two-day trip from Annapolis.

We stopped off at St. Petersburg, Florida, to view the progress of our new Morgan 34-foot sloop, *Sabrina II*. It was purchased in Annapolis after selling *Sabrina I*. We planned to sail it to Miami through the inland waterway and Lake Okeechobee as soon as it was complete.

The new residence exceeded our expectations in every way: three nicely appointed bedrooms, two bathrooms, and a two-car attached garage with screened-in pool and patio. It sat on a half-acre-corner fenced lot in a new subdivision, Green Hills. It was only a 15-minute drive (at that time) to the University of Miami campus. Unbelievably, its total cost on our arrival was $18,750. We were ecstatic! And the builder was proud of having everything in order for us when we arrived, with only one blip: a misplacement (by a foot) of the door from the master bedroom to the bath, so our king-size bed didn't fit! But when we showed up at his office the next morning, he promptly replied, "not to worry: I'll have it corrected by tomorrow afternoon." And he did! How's that for a warm welcome! Several weeks later we had an open house party to meet our new colleagues from the Campus, which drew about 50 guests (fortunately not all at the same hour) in our new and pleasing venue. It turned out to be a professionally challenging, but adventurous term in south Florida. We were off to a great start. The next seven years were marked by expansion and development of NHC, EML, and NHRL. It was also a time of melding with and participation in community activities; of acquiring new professional and other friends during visits and explorations throughout the Caribbean; and of professional liaison, conferences, and entertainment of international visitors interested in the work of NHRL and EML.

The Soviets' Visit

A year after our arrival at NHC, Bob White, head of NOAA, called me to ask me to receive a delegation of Soviet meteorologists, led by Academician Federov, at the National Hurricane Center and to conduct a small conference on tropical meteorology to introduce them to our operations. The delegation was scheduled to visit Washington, D.C., during its next stop.

During the visit, I immediately ran into trouble with the three *Komitet gosudarstvennoĭ bezopasnosti* (KGB) members monitoring the participation of the Soviet delegation. They continually and rudely interrupted their scientists (IN RUSSIAN) nearly every time they chose to speak. During lunch after the first session, I called Bob White to see if he had suggestions, since the conference was a travesty of oversupervision. His only words of caution were, "Don't rock the boat!"

However, before lunch was over, we decided on a scheme that might salvage at least some of the opportunity to develop a bit of collegiality. So, before the afternoon session began, Joanne and I invited the delegation to a picnic the next afternoon aboard *Sabrina II* while sailing to nearby Elliot Key, an uninhabited small island with a beautiful view of the Gulf Stream.

A few minutes before our departure the next afternoon (as I had confided earlier to Dr. Federov), I apologized to the group that I had been advised that *Sabrina II* could not legally take all of our group in the single boat. So, I asked my good sailing friend to bring his yacht to help transport the remaining members of the party. I asked Dr. Federov to identify the members of his group to go on each boat. He named the three KGB men and the most junior member of his staff to go on my friend's boat. I casually requested an equal number of junior members from my staff to accompany them on the second boat. The remaining senior scientists would sail aboard *Sabrina II*. As planned, there was insufficient time left to argue the assignment of staff before setting sail. All those aboard *Sabrina II* enjoyed the much welcomed freedom of speech and camaraderie. The tenseness from the scientific exchanges the first day was absent the final days of the Soviet visit, which continued in friendly communications afterward.

The National Hurricane Center, 1967–74

The return to south Florida in 1967 was a breath of fresh air for both Joanne and me—meteorologically, intellectually, and socially—especially to an NHC conveniently located on the campus of the University of Miami in a new five-story building, which housed NHC, NHRL, EML, the university's Department of Meteorology, and its computer science facilities. It was a structure tailored specifically to the activities housed there, including the daily weather map discussions, the collegial coffee hours, and the seminars sponsored by each unit. It was an idyllic venue for conducting cooperative programs of scientific research and operation—until 1979, when the machinations of the bureaucracy took its usual toll.

My tenure at NHC as its second director provided me with a nine-month overlap (at my request) with that of Gordon Dunn's. This allowed me to understand and discuss with Gordon the changes he had made when the Miami forecast office's duties were expanded to embrace hurricane-forecasting responsibilities for the entire nation. Dunn had achieved this with admirable ease and effectiveness. Also, this overlap allowed me time to become acquainted with the staff and to evaluate individually their apparent professional potentials or limitations.

A few months after the retirement of Dunn, I appointed what became known as the Senior Council, a five-person group that was considered the best equipped to review and advise me on the reorganization I considered essential for NHC. This council originally consisted of NHC's Deputy Director Arnold Sugg; the head of the Space Flight Support group Jesse Gulick and one of his forecasters, John Hope; the young, up-and-coming hurricane forecaster Neil Frank; and finally, my old friend and former student Gill Clark. The Senior Council took its responsibilities very seriously, arguing the merits and problems as a group for each change or new proposal. On the rare occasion when the final decision was counter to the consensus opinion, the council was unanimous in supporting it before the staff.

A Hurricane Encounter

Late spring of 1967 saw the graduation of Joanne's eldest son, David, from Yale and her younger son, Steven, from private school. We'd planned for six months to take them on a celebratory cruise though the Bahamas aboard *Sabrina II* along with their sister, Karen. Early September is high-risk season for hurricanes, factors well known by everyone at NHC. But not to worry, we were not going far, and if the conditions began to deteriorate, we could cancel our plans to leave or quickly return in ample time.

As departure time approached all but one of a string of easterly waves passed harmlessly into the Gulf of Mexico and the last gave no evidence of developing. So, we packed our provisions aboard and shoved off for the Bahamas under nearly cloudless skies at dusk, as we had numerous times before. But, as always, in the cockpit I had a slight feeling of wariness, especially during a nighttime passage. What I needed to be wary of was falling prey to overconfidence—and this time I did!

The stars began to disappear and the pleasant 12–15-knot east-northeast wind began to back (counterclockwise) and increase slightly, an ominous change, but with a few lights in the Bahamas in sight, it was too late to return to Florida, even with the advantage of tailwinds and our auxiliary motor.

As storm conditions steadily increased, I sent Joanne, Steven, and Karen into the cabin to secure the windows from rain and sea incursions, and themselves from jarring action from the waves. David helped me in the cockpit as we made our way to the nearest island and beat our way into the broad channel from the open sea, tacking back and forth to stay within the channel while the storm raged on. David read for me the changes in water depth, the compass reading, and the wind speeds while I wrestled with the helm and brought us about as needed to retain adequate water depth. All of this continued for two to three hours, after which the rain, stormy winds, and sea conditions gradually abated and we gratefully moved into the shelter of a private marina and rested. During our stormy ordeal, David had read wind values from our sloop's anemometer as high as 55 knots, with sustained values above 45 knots. I cannot speak for David or the three in the cabin, but strangely I don't recall any period during this ordeal when I was frightened. I probably was just too busy doing what I knew I had to do to be afraid.

What I later found out was that a low pressure storm had formed along a stalled front to the north. It intensified into a tropical depression as it moved slowly southwestward over the Bahamas—apparently north of our location—and then turned quickly to the northeast, clearing our skies.

Happily, the remaining nine days of this cruise were as pleasant and joyfully memorable as we had ever expected.

The tropical depression later became Hurricane Doria and made a curious track first out to sea and then back over the mid-Atlantic states, where it did minor damage but took three lives. Curiously, Doria followed a track and behavior pattern very similar to the controversial hurricane of 1947—controversial because the 1947 storm had been the subject of a seeding experiment that was part of Project Cirrus. Doria had similarly reversed its course but without the benefit of seeding.

Hurricane Camille

In the late 1960s, the satellite analysis group in Washington, D.C., had been given quite a bit of responsibility for making independent judgments in issuing hurricane warnings for international areas outside NHC responsibilities. So, a procedure had been established for coordination between NHC and the National Meteorological Center in Suitland, Maryland. In late August of 1969, a tropical storm formed off the western tip of Cuba and was named Camille. We had had some rather lively arguments between NHC and Suitland on two successive advisory cycles about changes of intensity in Camille as it moved into the Gulf of Mexico. The staff in Miami, both at NHRL and NHC, were of one mind that the storm had vastly strengthened and that the shrinking of the central dense overcast (CDO) was the result of sinking of air transported up in the eyewall in such abundance that it could not be carried away from the core rapidly enough. The satellite analysis unit thought the shrinking CDO meant the storm was weakening. Finally, I decided to stop arguing and stated categorically in the next advisory that Camille was becoming a very severe hurricane. But we were strapped for recon aircraft to verify our analytical conclusions while the satellite analysis people continued to snipe at us.

At the time, there were two hurricanes in progress, the other being Debbie, which was north of Puerto Rico and the subject of a STORMFURY experiment. Nearly all the Navy aircraft had deployed to Puerto Rico for this exercise and only two remained in Jacksonville. The first Navy plane sent to penetrate Camille had to turn back with engine trouble. The second got off several hours later, but in approaching Camille it decided it was too severe to penetrate and turned back without getting any quantitative information from the storm core. To complicate matters further, the only Air Force recon planes available at that time were on the West Coast. With a crisis at hand, I called Scott Air Force Base near Belleville, Illinois, to talk to the Air Weather Service commander personally, and asked what could be done to get a reconnaissance plane into Camille. He replied that he would have to fly one from the West Coast but would do the best he could. So, a plane with two crews was flown to Tampa, where it was refueled and dispatched immediately into Camille, using the alternate crew. When they arrived, they reported a central pressure of 901 millibars (later corrected to 905). Our worst fears were confirmed, and we pulled out all stops to make it known that the worst storm of record was descending on the central Gulf Coast.

More than 75,000 people were evacuated from the Mississippi coastal area that was most heavily beset by Camille. If the 12–15 hours of explicit warning of a record storm had been later in coming, or if evacuation authorities had been less efficient in conducting the evacuation, then it has been estimated that thousands of additional lives might have been lost.

The center of the hurricane passed over the Mississippi coast near the towns of Clermont Harbor, Waveland, and Bay St. Louis about 11:30 p.m. (CDT) on Sunday, August 17. Maximum winds near the coastline could not be measured, but from an appraisal of the character of the splintering of structures within a few hundred yards of the coast made by Herb Saffir, velocities probably approached 175 knots. The highest recorded storm surge observed, which apparently occurred near Pass Christian, Mississippi, was measured at 24.6 feet, higher than any previous storm tide of record.

I traveled from NHC in Miami to the Camille damage scene a scant 24 hours after the landfall and spent two days surveying the damage in the Air Force helicopter and talking with survivors. I wrote about what I witnessed in the annual hurricane season summary published in the *Monthly Weather Review*:

> The old ante bellum residences, which had stood in grandeur along the Mississippi coastline from Pass Christian to Biloxi and had withstood the ravages of many hurricanes for more than a hundred years, had been totally or substantially destroyed, with few exceptions. In the Pass Christian–Long Beach area, little trace of intact structures such as roofs or wall sections was observed within 100 yd of the coastline. Here, houses had been swept completely off their foundations and splintered into unrecognizable small pieces, characteristic of the wind damage ordinarily associated with major tornadoes.

I found myself at New Orleans International Airport boarding a plane to return to Miami when I was handed an urgent message to call Washington, D.C. When I did, someone in ESSA Administrator Bob White's office told me to return to Biloxi, Mississippi, and accompany Vice President Spiro Agnew on a survey of the damage area. Five minutes later I would have been on my way to Miami, there would have been five—not six—reports given to Agnew, and goodness only knows what the outcome would have been about hurricane reconnaissance facilities. Nevertheless, I joined Agnew on Air Force One and pointed out for him the most significant evidence of damage from flooding and from wind that we had surveyed earlier. Back at Biloxi, Agnew asked for summary statements from the emergency warning people, disaster relief officials, the American Red Cross, and the NHC. In my statement I recalled the compounding of uncertainties when Camille was rapidly becoming a record storm when the satellite analysis group in Suitland had insisted that the storm was losing strength, while the NHC forecasters concluded that it was strengthening. And without adequate reconnaissance, we couldn't be certain how loud to ring the bell for evacuation.

Agnew wanted to know what the problem was in obtaining the kind of recon data we needed to verify the strength of the hurricane. My response was what lit the fire that enveloped both ESSA and the Department of Defense and, I was told, almost got Bob White fired. I said both the Air Force and the Navy had been trying to get a better instrument and recorder system aboard their planes, and the Navy had been trying for years to replace their antiquated Super Connies, whose performance in severe hurricanes was questionable. I continued, stating that every effort in the last three budget cycles was frustrated because the administration turned down the budget requests for these much-needed facilities.

This placed the forecasters and the warning systems in great jeopardy on this occasion. Then, on sudden impulse, I lit the fuse, adding that, personally, as a U.S. citizen, I felt it was high time somebody got on the ball and saw that this problem got proper attention in the right places in Washington.

I knew I was taking a great risk. I had been around long enough to know I was making statements a civil servant had no business making if he valued his career. But I saw the opportunity and decided in a split second to make this risky personal comment.

The acutely painful result was that Agnew flew back to Washington, turned right around and flew to the "little White House" at San Clemente, California, and briefed Nixon. Nixon, in turn, called Bob White to California along with the responsible officers in the Air Force and the Navy. They all got thoroughly chewed out. Then, in turn, I got promptly called to Washington and was chewed out by Bob White. I believe Bob really thought his job and career were placed in tremendous jeopardy, and it probably was, though not intentionally. But it shows how things work within the bureaucracy. This time, fortunately, it all turned out well. No one got fired or lost a career, the Navy got their P3s, the Air Force got millions for new instrument systems, and from all the momentum generated, ESSA was able to improve its own research aircraft facilities.

The Saffir–Simpson Hurricane Scale

Camille pointed out to me a dire need to be able to easily communicate the dangers of a looming hurricane to emergency managers. Luckily, my friend Herbert Saffir had already been working on a windstorm scale for the World Meteorological Organization. He divided the winds from hurricane force to 150 miles per hour into five equal categories in order to evaluate wind damage.

I borrowed what he'd proposed and the NHC staff tweaked some of the levels (including raising the upper bound to 155 miles per hour), correlated central pressure vales for each category, and added storm surge estimates. In 1974 we placed this scale in operational use at NHC. At the 1975 Interdepartmental Hurricane Conference, it was officially dubbed the Saffir–Simpson Hurricane Damage Potential Scale, an action to which neither Herb Saffir nor I were privy. This designation was by committee action, not by our request, and a bit of an embarrassment to me.

Sabrina III

Eventually, *Sabrina II* was replaced by *Sabrina III*, a beautifully equipped Gulfstar 43-foot sloop. It was our first boat with a powerful diesel auxiliary engine capable of motor sailing us home from the Bahamas at an 11-knot pace (wind and seas willing). To accommodate our acquisition, we purchased a Coral Gables property, a three-quarter-acre pie-shaped lot at the dead end of Gallardo Street in Old Cutler Bay, Florida. It was situated on a canal two blocks inland from the Biscayne Bay waterfront. For this unique setting, we designed and contracted the construction of our dream house with docking facilities for *Sabrina III* in the rear.

The Caribbean

During our time in Miami, we made numerous trips around the Caribbean, both for professional and personal reasons. We spent time on virtually every island and the bordering countries of Central America. The professional trips were on research flights, carrying out studies on tropical meteorology, including hurricanes or giving lectures on various means of survival and protection from the annual parade of tropical storms in that region. On these missions we met and developed lasting friendships among meteorologists in the area.

The personal visits were on board the *Sabrina III*. Paul and Betty Agnew, sailing shipmates from the University of Miami, joined us on many of our sailing adventures in these areas, as did both Peggy and Lynn and their friends. The island of Saint Vincent was particularly memorable for its remarkable view of the green flash during sunset, and for our exploratory ascent of the volcano at the center of this picturesque tiny island.

When we arrived at the summit we were disappointed, but not surprised, to find clouds obscured the view of the caldera bottom lush with coconut palms. But what followed minutes later really captured our attention. A small group of young Sunday school students arrived. Their leader quickly shushed their shouts of disappointment and promised to do something about the clouded scene. With little hesitation he offered a brief but eloquent prayer that the clouds be removed long enough for the children to feast on the beauty of what lay below. Sure enough—for a period of about 10 minutes the clouds disappeared, revealing a crisp sunlit view of this unique summit, the caldera, and the island at its bottom! The two meteorologists in our group of climbers remained dutifully silent in response to the children's shouts of wonderment that they had truly witnessed a miracle that Sunday morning.

Retirement from NOAA

By 1973, however, both Joanne and I were convinced we needed to seek a different venue for our final years in meteorology. In particular, Joanne was unhappy with NOAA's increasing pressure to move her weather modification experiments into operational status. For my part, I had been at NHC past the five years I usually invested in any particular position within the government. To add insult to injury, supervisory and fiscal management of NHC was transferred from headquarters to the Weather Service's Fort Worth, Texas, regional office. I now spent 10 times the effort to carry out the same administrative tasks. The move, needless to say, did not improve the morale or overall performance of personnel at NHC.

It was then that Michael Garstang, an old friend from Joanne's Woods Hole Oceanographic Institition days, became instrumental in obtaining an offer for us both to join him on the faculty of a new environmental science department being established at the University of Virginia. Joanne was offered an endowed chair and I a position on "soft money." I announced my intention to step down as NHC director in 1974, affording Neil Frank a year's transition to follow me into the post.

The operational challenges at NHC and the Miami area had provided us with many gratifying experiences, scientific explorations, happy encounters, and lasting friendships, which enriched our lives and sense of achievement. Among my accomplishments were the inauguration and success of the Saffir–Simpson scale, the concept of in situ evacuation during hurricane emergencies, the entrainment of societal experts from academia into designing more effective means of communicating the urgency to respond to emergency warnings, a new system of small craft weather warnings, and strongly supporting the annual conferences sponsored by Gilbert White at the University of Colorado Boulder to study means of protection from hurricanes. In addition, we acquired new friends during frequent recreational sailing expeditions in the Florida Keys, the Bahamas, and the Caribbean. We learned many lessons about the weather while sailing that were not available in textbooks or lectures in tropical meteorology.

Chapter 9
The University of Virginia, Simpson Weather Associates, and Retirement

University of Virginia

Joanne and I arrived in Charlottesville, Virginia, during the spring of 1974, enthused with the prospects of setting out on a new career in a magnificent and historic academic setting still bristling with the spirit and memory of Thomas Jefferson. It was a move dampened only by the loss of our memorable weekly sailing expeditions in the Gulf Stream waters off south Florida and the Bahamas.

Joanne was bubbling with excitement, not only at the prospects of the endowed chair she would occupy, but also the opportunity to bring with her Roger Pielke, the brightest and most promising young scientist from her NOAA research group in Miami.

WMO's Conference on Tropical Meteorology in Nairobi

Just after our retirement from the government, but before our move to the university, Joanne's international commitments brought about our lengthiest travel in 1974, primarily to participate in the WMO International Tropical Meteorology Meeting in Nairobi, Kenya, but it ended up extending much further.

Our first stop was Israel, where Joanne gave a few lectures in Jerusalem. Our departure was interrupted by a traffic-stalling ten-inch snow, resulting in two extra nights in the King David Hotel. During the holdover, Abe Gagin flew us over the area of his weather modification experiment, and over historic Masada. When the roads finally cleared, we drove a rental car over most of Israel, ending up at Kibbutz Nir David, one of the oldest dating back to the 1930s. One of our professor friends from Jerusalem had prearranged for us to be guests for a long weekend. It was an amazing and educational experience.

We finally reached Nairobi after having to backtrack to Athens, Greece, before we were allowed to fly from Israel to Nairobi. The WMO conference, however, was not the most rewarding event of the journey. Afterward, we joined Mike and Elsabe Garstang who, from their long early life experiences and familiarity with Africa, had made detailed arrangements for the four of us to enjoy well-planned safaris. From three prime venues in Kenya and Tanzania, we had the experience of a lifetime observing the wildlife of central Africa, including Ambeseli, the Serengeti, and Ngorgora Crater. It was something Mike Garstang described admirably, and skillfully illustrated with his wildlife sketches after his retirement from the University of Virginia (UVA).

South Africa

Leaving the venues surrounding Nairobi, we traveled to South Africa, first its capital Pretoria and afterward to Nelspruit. This would be the site of Simpson Weather Associates' weather modification experiments in a few years. This would be a proof-of-concept undertaking, exploring hail suppression through seeding. Joanne would be joined by an old colleague, David Atlas, in going up in the observation jet that plunged down on any targeted cloud.

Finally, we went to Johannesburg, South Africa's largest city where, on our last day there, we had an unexpected encounter. After an interesting evening dining in the rotating skyscraper restaurant in midtown, as we arose to leave, we were hailed from the opposite side of the dining room by an old friend and colleague, Glenn Brier and his wife. Unbeknownst to us, they had traveled to South Africa on vacation. It was an improbable location, halfway around the world from home, for an exuberant extension of our evening at this unlikely spot. We exchanged accounts of our quite different ports of call en route.

The next day, Joanne and I completed our fascinating journey and explorations of Africa by visiting magnificent Capetown. We then flew to Rio de Janeiro, Brazil, where Joanne had agreed to consult with a private corporation on a water resources problem. We arrived in Rio with the city abuzz with plans for the annual Carnaval festivities. Our sponsor, who met us at the airport near midnight, was unimpressed by our pleas of travel weariness. Without hesitation, he hustled us off to a remote countryside location where (what we estimated to be) a thousand or more participants were riotously engaged in auditing and selecting theme songs for Carnaval.

By this time, however, our exhaustion prevailed. After excusing ourselves, we sought a cab and hurried to our hotel. Forty-eight hours later, we were on our way home for a short-lived respite before busying ourselves with the move to Charlottesville and its academic environment.

Disappointment with UVA

For me the adjustment from the busy life and professional activity I had enjoyed at NHC, with an eager staff ready to meet the challenges we continually faced there, to the quiet life of a faculty member at UVA was a hard one. While I found the physical environment delightful to explore, neither Joanne nor I ever felt at home on the UVA campus or with the new department to which we were attached.

Despite the verve and strong support of Mike Garstang, who had succeeded in bringing us to UVA, the zest and excitement that Joanne had brought to her new position soon faded away. She quickly was made to realize that as a woman, she at best was unwanted. Her attempts to participate in faculty affairs as a chaired professor were in fact, frequently resented, a resentment that I found myself returning in kind on her behalf. She's just not of that temperament, but she was terribly hurt and developed a deep-seated depression for which she required extended treatment. Obviously, I was disturbed and suffered along with Joanne over the rude and crude treatment she was receiving on campus, especially from the old guard.

Despite all this, I could not help becoming quite fond of the broader Albemarle County environment, although the attraction faded quickly on Mr. Jefferson's famous campus. Off campus we continually enjoyed the friendship and social attentions of a limited number of faculty families, especially from the Garstangs. But both of us turned our professional interests and efforts to off-campus activities, Joanne to international programs (such as the Global Atmospheric Research Program [GARP] Atlantic Tropical Experiment [GATE] and the Winter Monsoon Experiment [MONEX]) and to national committee work and I to numerous requests for consulting services.

Simpson Weather Associates

Since my retirement from government, I had received many request for consultation on weather-related projects. I solved the demand by establishing and incorporating Simpson Weather Associates (SWA). We began by enlisting UVA faculty members to assist with technical services on a "when actively required" basis.

But even with these changes in professional activities, it didn't take long to conclude that we must seek still another venue as an outlet for our professional pursuits. This began almost immediately by planning foreign travel to join colleagues elsewhere with common research interests.

The GATE Expedition

Our first year at the University of Virginia, Joanne and I accompanied many of our colleagues, some from overseas, to participate in what became the largest meteorological expedition ever to study tropical meteorology and the origin of hurricanes. It was called GATE.

Based in Dakar, Senegal, GATE employed four oceanographic research vessels and a dozen or more international research aircraft (including NOAA's C-130 and one of its DC-6s). Near the airport were its data collection facilities, where daily map discussion helped with the planning of that day's scientific experiments.

Joanne participated in low-level boundary layer missions, rejoining her old colleagues flying on the NOAA C-130, gathering vital cloud information. I collected data from a NASA C-135 at the 35,000-foot level. These missions were carefully planned, coordinated, and conducted. Most of the flights were routine in nature; the greatest cause for excitement were the enthusiastic conversations occurring after landings while the scientists and flight staff walked the half mile to a restaurant, comparing their separate experiences during the flights.

The scientific data were later analyzed and most were published in dozens of research papers, and they produced new concepts and understanding of the role of the tropics in the general circulation and the generation of hurricanes and other severe storms. The data also generated many collaborative research programs.

Typhoon Moderation and the Philippines

In the early 1970s, the Philippines were struck by a series of devastating typhoons. Philippine First Lady Imelda Marcos was greatly saddened by the resulting enormous loss of life and property, and she hunted for a means of controlling or reducing her nation's annual losses from tropical cyclones. Her search began with a notable friend in Washington, D.C., who put her in touch with an entrepreneur in Falls Church, Virginia. I'm sorry that I can't recall his name, but ultimately he contacted SWA and me. When he first visited me in Charlottesville, he made a bold pitch for us to jump aboard his scheme to offer the Philippines a STORMFURY-style program managed by SWA. It was to be called typhoon moderation (TYMOD).

I struck a very discouraging note, mentioning a probable cost of many millions of dollars and the need to conduct a proof of concept experiment prior to any operational attempts to control typhoons. These experiments wouldn't rule out a negative result that would preclude any operational phase, although they would produce valuable research results either way.

Despite my pessimistic evaluation, arrangements were made for me to travel to Manila, Philippines, to discuss with Imelda and her Philippine experts the nature of the project. I did, was royally entertained, and to my surprise we were contracted to prepare a working plan for a 10-year proof of concept experiment and including a best estimate of total costs of the program.

After days of consulting with a select panel of senior scientists, we prepared a skeleton mock-up of such a program, which I presented to Imelda Marcos at her Waldorf Astoria quarters in New York City, several months later. She seemed to be pleased with the presentation and in response to my query whether the costs would pose a problem, she brushed it off, indicating she had already discussed it with her husband and it would be presented to his staff when she returned home.

Unfortunately, this optimistic outlook rapidly deteriorated as President Marco's regime fell into disrepute and the First Lady's ambitions on behalf of her typhoon-beset people quickly ended. However, many of us involved continued to admire her motivation and stubborn determination regarding the welfare of her people.

England and the London Conference

In the summer of 1978, Joanne was invited to be a distinguished lecturer in tropical meteorology at a North Atlantic Treaty Organization (NATO) training symposia for young postdoctoral meteorologists in London. Joanne was able to renew her acquaintances with colleagues she'd acquired during a year's sabbatical in Great Britain years earlier.

However, for us, the highlight of the trip occurred after the symposium. By now, Lynn had a teaching job in Germany. She was able to take a 10-day leave and join us for a journey northward to the Simpson ancestral home in Scotland, visiting en route many ancient cathedrals, historic sites, and monuments that Joanne was familiar with, but were new to Lynn and me. We traveled up the east coast of Great Britain to Inverness, Scotland, and back down the west coast to London. The Inverness area was not only the ancestral home of the Simpson family but of their Fraser clan and the disastrous ancient battlefield of Culloden. On our return trip along the west coast, we spent the night in Oban, Scotland, attending a festival in which old Scottish music and dancing were performed. It was one of the most memorable opportunities the three of us had had to spend time together alone.

The frosting on the cake came when we returned to London during the Gilbert and Sullivan festival. We were enchanted by a performance of *Yeomen of the Guard* performed within the walls of the famous Tower Green, all adorned from atop the walls with music from long, golden heralding trumpets.

Joanne's Rescue

Ultimately in 1979, David Atlas, our longtime professional friend, was able to "rescue" Joanne from UVA, offering her a permanent government leadership position in the Earth Sciences group at the National Aeronautics and Space Administration (NASA)'s Goddard Space Flight Center. Here she rapidly acquired distinctions for her leadership, and was appointed Goddard Senior Fellow. Once again we were returned to the Washington, D.C., area. The management of SWA remained in Charlottesville, prospering under the astute leadership of Mike Garstang, and a short time later with David Emmitt. My continuing participation consisted primarily of weekly commutes from the Washington area.

Joanne's duties at Goddard were in directing the Tropical Rainfall Measuring Mission (TRMM) project. This involved a low-orbiting satellite equipped with a downward-scanning radar that orbited preferentially over the tropics. Over the life of the project, it created a detailed catalog of tropical rainfall, including many hurricanes. She oversaw this valuable program for over a decade ,and it provided a fitting capstone to her government career.

Cruising Southeast Asia, the Middle East, and Mediterranean

After our retirement, Joanne and I shared dozens of cruises, travel experiences, and explorations. Many do not appear in this account, not because they were unimportant or uninteresting, but because they did not contribute sufficiently to the purposes intended here. Nevertheless, our final ocean cruise on Renaissance Cruises' humongous vessel, the *R One*, was sufficiently unique to warrant inclusion here, mainly because it was marred by international tensions and uncertainties, which resulted in the cancellaton of so many of its planned ports of call.

The *R One*, the flagship of the R-Class ships, was not chosen because of its size (from our experience, bigger is not better when it comes to ocean cruising) but its itinerary was attractive when booked, though less so at boarding time. Its desolate port of entry was an hour's bus ride from Bangkok, Thailand's capital. When approaching it, *R One* appeared a stubby ugly duckling; however, once aboard it proved attractive and very comfortably equipped. We enjoyed the accommodation while aboard, although not nearly as much as our previous Renaissance experience in the Baltic Sea aboard the much smaller *R Seven*, with its spacious accommodations, including such luxuries as large walk-in closets.

The trip southeastward to Ho Chi Minh City, formerly Saigon, South Vietnam, and then southward to Singapore was a study in contrasts. After encircling the Indian Ocean to the north and west and crossing the Arabian Sea, we entered the Persian Gulf. This area overall was a disappointment, primarily because of the military and political unrest and uncertainties, and the cancellation of scheduled ports of call not only in the Persian Gulf but the Rea Sea, Sinai, and Egypt. Memories of this tended to sully the pleasant and welcoming experiences in the few oil-rich United Arab Emirates we were allowed to visit, especially that of the Sultanate of Oman. Such a breath of fresh air after the exposure to other wealthier nations of the area! It was less constrained by religious mandates, and governance policies. Likewise, it was like emerging into another world as we exited the Suez Canal into the eastern Mediterranean, and the termination of our cruise at the port of Piraeus, Greece.

Reflections on Science and Religion

During my 100 years from childhood to maturity and the declining years, I've reflected on this remarkable and varied journey. I am occasionally confronted with the need to reexamine my early background, and the experiences and motivations that most influenced my thought processes and guided my decisions. My conclusion has always been the same: my childhood as a member of a deeply religious family whose lives were primarily influenced and guided by the Christian ethic, the discipline of the Methodist Church, and their abiding faith in Christianity. The religious instruction of the children was by example with occasional penetrative questioning of misbehavior but never the carrot-and-stick approach. With me this worked very well, motivating the better attributes of my nature. On occasion, as was obvious to me, I was a disappointment to the family, although it rarely resulted in a scolding. Through it all, their example, not always understood nor at the time welcomed, left an indelible mark on my ability to reason and to respond positively to life's challenges.

The challenges rarely had their origins specifically in the context of religion. They usually involved around-the-table discussions of current events during mealtimes, often posing problems whose resolution was the subject of comment and exchange of views. Sometimes it revolved around human behavioral problems and the distinction between acceptable and unacceptable behavior, individually or as a group. We also discussed more appropriate alternatives as well as corrective measures if necessary.

I was well into my college years before I was able to recognize how much these early family experiences had influenced my ability to accept and live effectively with the challenges of life during the Great Depression, when most families—including my own—suffered debilitating hardships. Oddly, however, the hardships did not register with me as more than temporary setbacks or bad luck. To me it was an exhilarating period, a first time away from home to face life, making my own decisions. I had to make my way in the world alone with interesting challenges, including short careers in music, in architecture, and a career in academia. Later, I pondered the sustained good fortunes I had enjoyed during college and early professional years on my own, by comparison with the experiences and opportunities of many of my friends and close associates. It became clear that my experiences were no accident but in large measure were due to the guidance and sustained impact during early youth as a cherished member of my Christian family circle.

Sadly, as I belatedly realized, I'm afraid I never succeeded in effectively communicating the depth of my appreciation of my parents' contribution in molding my Christian faith, its underlying imperatives of ethical behavior, and my sense of fulfillment and optimistic outlook on life. It was an appreciation my mom and dad so richly deserved.

Through my university years into my professional life, my Christian faith remained steadfast, although enlarged in some respects from exposure and experience in scientific research, which requires reason in search of understanding. At Southwestern University, Dr. Gray, professor of religion and a Methodist minister, was the first person I encountered who expounded the need to undergird religious faith with reason. In doing so, he referred to many biblical accounts that appeared to be in conflict, even contradictory. He did not consider this a challenge to his basic faith and provided reasoning why. Gray was not the only one who cogently connected faith and reason, which reordered a number of my reasoning processes, both professionally and in shaping my religious faith.

In more advanced studies in science, particularly in my studies of meteorology, hurricanes, and other severe storms, I soon gained an understanding of the scientific method applied to research to gain a fuller understanding of physical processes. This involved reasoning based on what was already known in designing a hypothesis to be tested, forming the framework for research toward a fuller understanding of unexplained observations or of the physics. Curiously, during church (where I was a regular attendant through midlife), I soon found that my attention at services was occupied almost exclusively with 1) the religious music, which I love; and 2) meditation, pondering broadly the more recent challenges and puzzles of life while shutting out the message of the day from the pulpit. As I now recall, of the 17 Methodist churches in which I have held memberships, only 4 out of a total of 21 ministers regularly captured my attention with their sermons. Nevertheless, through it all I consider myself a Christian, with acceptance of Christian beliefs that I think I understand, and faith that what I don't understand now will someday be illuminated by a more sophisticated acquisition of knowledge.

My perception of religious faith is the acceptance of those details or tenets of religion that are beyond my present ability to understand, based on present levels of knowledge and reasoning, but bolstered by a confidence that as knowledge expands and more informed reasoning leads to fuller understanding, the scope of religious elements needing acceptance by faith will progressively diminish.

In many ways this process resembles that by which physical research leads to the expansion of knowledge in basic physics, including breakthroughs in understanding. However, historically, physical understanding tends to be acquired incrementally: no single breakthrough quite gets it right but often ends up pointing the way to succeeding steps toward understanding. This outcome has occurred even in the profound contributions by the most eminent scientists: from Galileo to Isaac Newton and Albert Einstein. In my meditations it occurred to me that religious faith as a link to understanding bears a remarkable similarity to the hypothesis as applied to the scientific method in furtherance of scientific understanding.

This has been particularly evident in the efforts both Joanne and I experienced in understanding the behavior of hurricanes. For example, in the 1940s, it was commonly accepted that both the growth and steering of hurricanes was entirely explainable by the changes occurring in the larger-scale environment, independent of the hurricane circulation itself. Nearly half a century later, further research and supporting reasoning made it clear that many aspects of hurricane behavior depend on the interactions between successively smaller scales of motion within the hurricane and its immediate environment. Yet, there remain so many gaps in understanding hurricane behavior. In religion the linkage between faith supported by reasoning is a concept that strengthened my religious faith and my willingness and comfort in conceding, "I don't fully understand, YET, but I hope and believe someday I will."

Having acknowledged and enlarged the impact of my Christian family during my early years, I must further acknowledge that they were not the only factors importantly influencing decisions that challenged my early Christian background and training, namely, marriage: a first marriage to Mazie Houston, a devout Christian and devoted mother—a marriage that after 12 years had to be judged a failure in most respects except for our two wonderful children, Peggy and Lynn. It was a marriage I sought to terminate because of my conviction that otherwise my promising career as a contributing scientist would be stultified—if not destroyed—by continuing in it, a conclusion that I still believe was correct. But at best it was nevertheless a selfish one, which ended up costing me dearly in self-esteem and guilt but not in loss of my Christian faith.

Ironically 16 years later, and after two additional personal missteps, I married Joanne Malkus, a brilliant young scientist who shared common interests with me in tropical meteorology and severe storms. During our 45 years of marriage, we shared personal and emotional interests that generated lasting happiness and a sense of fulfillment, contributing to the maturity and professional growth of us both, reflected in wide recognition.

Curiously, in contrast to my happy family background, during her childhood, Joanne had been made to feel unwanted. Her sole exposure to religion was mainly through her grandparents and the private school (Buckingham Browne and Nichols, in Cambridge, Massachusetts) she attended as a child. While not an atheist, she found it difficult to acquire faith in Christianity, even though she participated with me in various church activities. She maintained a relentless thirst for truth in all walks of life, and was widely read in classical and religious literature. We had no difficulty in discussing our differing points of view on religion, and fully agreed that most public education suffers from a lack of inclusion of biblical literature at least as part of the history curricula. Our lack of agreement on religious issues was in distinguishing the logical continuity between faith and reason. It posed no threat to our well-established relationship. In fact, it forced us both to confess, "I don't know," but with differing viewpoints: one adding the clause "YET, but I have faith in its validity if only as an icon of truth"; the other, a qualifying clause implying "... and it's too intangible to pursue further."

Nevertheless, we both admired so much the enlightenment that exists in journalist Bill Moyer's interviews with scholars on the subject of faith and reason. Moreover, we both deplored the infectious sectarian militancy and violence under the guise of religion, now increasingly displayed the world over. In my view there is ample logic and reason to disarm these concerns by a thoughtful viewing of Moyers' TV series *Bill Moyers on Faith & Reason*, as it evenhandedly probes various views and concerns. His series, first appearing on the Public Broadcasting Service (PBS) network, is currently available on eight DVDs. It contains interviews with 12 scholars with widely differing points of views relating to faith and reason. I particularly recommend Mexican author Dr. Richard Rodriguez and Britain's Sir John Houghton, both in episode 5.

Finally, in reading the *New York Times* Science section on November 21, 2006, I found it puzzling to find such a preponderance of exclusionist views on science and religion expressed by so many notable scientists without recourse to caveats regarding the alleged adverse effect of science on religion. In science, understanding is continually advanced by well-conceived hypotheses for research.

Yet, as stated above, even the most important breakthroughs never quite get it right. The ultimate truth always lies ahead. Similarly, in religion ultimate understanding has always been a work in progress bound up in faith and reason: faith, the acceptance of religious details with the caveat, "I don't know YET, but will seek understanding"; reason, the struggle FOR understanding. Understanding may have to remain elusive for years before sufficient enlightenment is within grasp. Nevertheless, there is certainly enough strength and credibility in both science and most religions for intelligent proponents of one to avoid alarm about the adverse effect from proponents of the other.

Bob and Joanne Simpson at Roosevelt Roads Naval Station,
participating in Project STORMFURY

Bob and Joanne Simpson,
with Herb Riehl.

Bob and Joanne Simpson on
their wedding day, January 1965.

Bob Simpson, as director of the National Hurricane Center
in the main forecast room; Coral Gables, FL, March 1970.

Bob Simpson and Herbert Saffir, collaborators in developing
the Hurricane Damage Potential Scale, in Spring 2001.

Bob and Joanne simpson at a Cosmos Club party,
launching Bob's book *Hurricane: Coping With Disaster,*
December 2003.

Forrest M. Mims III with Bob Simpson. Mims dedicated his book, *Hawaii's Mauna Loa Observatory,* to Bob. Photo by Barbara Schoeberl.

Bob Simpson at the 2012 Hurricane and Tropical Meteorology Conference, in Ponte Vedra Beach, FL. Photo by Neal Dorst.

Robert H. Simpson's Career Positions

Year	Position	Place
1929	Residential Design Assistant (Le. Fite Corp)	San Antonio, Texas
1930–33	Undergraduate Student of Math & Physics	Southwestern University Georgetown, Texas
1933–35	Graduate Student in Math & Physics	Emory University Atlanta, Georgia
1935–40	Public School Teacher	Crocket, Ft. Stockton, and Corpus Christi, Texas
1940–41	United States Weather Bureau Weather observer	Brownsville, Texas
1941–42	Weather observer and Facility Manager	Swan Island, West Indies
1942–43	Hurricane Forecaster	New Orleans, Louisiana
1943	Hurricane Forecaster	Miami, Florida
1943–44	Graduate Student of Meteorology	University of Chicago Chicago, Illinois
1944–45	Air Force Instructor Tropical Meteorology	Howard Air Force Base Panama
1946–48	Professional Assistant Office of Department Chief	U.S. Weather Bureau HQ Washington, D.C.
1948–52	Meteorologist In Charge (MIC)	Honolulu Hawaii & Pacific Area
1952–55	Professional Assistant Office of Department Chief	U.S. Weather Bureau HQ Washington, D.C.
1956–59	Director, National Hurricane Research Project	Morrison Air Field West Palm Beach, Florida
1960–62	PhD Candidate in Meteorology	University of Chicago Chicago, Illinois
1962–64	Deputy Director of Research (Severe Storms)	U.S. Weather Bureau HQ Washington, D.C.
1962–63	Director, Project STORMFURY	Miami, Florida
1964–67	Director of operations U.S. Weather Bureau	U.S. Weather Bureau HQ Washington, D.C.
1965	Certified Consulting Meteorologist	American Meteorological Society

continued >

Robert H. Simpson's Career Positions (continued)

1967	Deputy Director National Hurricane Center	Coral Gables, Florida
1968–73	Director National Hurricane Center	Coral Gables, Florida
1974	retired from U.S. Government service	
1974–79	Professor of environmental Science	University of Virginia Charlottesville, Virginia
1979	Founding Director Simpson Weather Association	Charlottesville, Virginia
2000	Director Emeritus Simpson Weather Association	Charlottesville, Virginia

Robert H. Simpson's Major Career Achievements

Year	Position	Place
1947	First "Piggyback" Research Flight U.S. Air Force WB-29	Bermuda
1950	Inaugurate Observatory Facility	Mauna Loa Summit Hawaii
1951	Typhoon Marge Research Flight	Andersen Field Guam
1952	Negotiated Establishment of the Texas Radar Network	Austin, Texas
1955	Negotiated Funding for the Mauna Loa Slope Observatory	Washington, D. C.
1963	Establishment of the National Severe Storms Project (NSSP)	Norman, Oklahoma
1968	Reorganize National Hurricane Center	Coral Gables, FL
1968	Begin Annual NHC Visitation of the Caribbean Weather Stations	Coral Gables, FL
1968–69	Revised Hurricane Warning Procedures	Coral Gables, FL
1970–74	Developed Saffir-Simpson Scale	Coral Gables, FL
1974–78	Participated in Four International Weather Expeditions	Dakar, So. Pacific and Australia
1978	Develop Project Tymod (typhoon modification) for the Philippine Government	Charlottesville, VA

Publications

Formal Publications and Proceedings
of Congress and International Symposia

Gentry, R. C., and R. H. Simpson, 1956: Hurricanes. Annual report of the Board of Regents of the Smithsonian Institution, Smithsonian Institution Publ. 4272, U.S. Government Printing Office, 301–328.

Malkus, J. S., and R. H. Simpson, 1964: Modification experiments on tropical cumulus clouds. *Science*, **145**, 541–548, doi:10.1126/science.145.3632.541.

Malkus, J. S., and R. H. Simpson, 1964: Note on the potentialities of cumulonimbus and hurricane seeding experiments. *J Appl. Meteor.*, **3**, 470–475, doi10.1175/1520-0450(1964)003 <0470:NOTPOC>2.0.CO;2.

Namias, J., G. Dunn, and R. H. Simpson, 1955: A survey of the hurricane problem. *Trans. N.Y. Acad. Sci.*, **17**, 346–351.

Rodgers, Golden and Halpern, Simpson Weather Associates, and H. W. Lochner, Inc., 1981: Hurricane evacuation and hazard mitigation study. Rep. for City of Sanibel, FL.

Simpson, J., and R. H. Simpson, 1975: On the structure and organization of clouds in the GATE area. Preliminary scientific results of the GARP Atlantic Tropical Experiment, GATE Rep. 14, Vol. II, WMO and International Council of Scientific Unions, 160–167.

Simpson, J., R. H. Simpson, D. A. Andrews, and M. A. Eaton, 1965: Experimental cumulus dynamics. *Rev. Geophys.*, **3**, 387–431, doi:10.1029/RG003i003p00387.

Simpson, J., R. H. Simpson, J. R. Stinson, and J. W. Kidd, 1966: Stormfury cumulus experiments: Preliminary results 1965. *J Appl. Meteor.*, **5**, 521–525, doi:10.1175/1520-0450 (1966)005<0521:SCEPR>2.0.CO;2.

Simpson, J., G. W. Brier, and R. H. Simpson, 1967: Stormfury cumulus seeding experiment 1965: Statistical analysis and main results. *J. Atmos. Sci.*, **24**, 508–521, doi:10.1175/1520-0469(1967)024<0508:SCSESA>2.0.CO;2.

Simpson, J., Th. D. Keenan, B. Ferrier, R. H. Simpson, and G. J. Holland, 1993: Cumulus mergers in the Maritime Continent region. *Meteor. Atmos. Phys.*, **51**, 73–99.

Simpson, R. H., 1944: Subtropical rain showers from stable cloud forms. *Bull. Amer. Meteor. Soc.*, **25**, 367.

Simpson, R. H., 1946: On the movement of tropical cyclones. *Eos, Trans. Amer. Geophys. Union*, **27**, 641–655, doi:10.1029/TR027i005p00641.

Simpson, R. H., 1947: A note on the movement and structure of the Florida hurricane of October 1946. *Mon. Wea. Rev.*, **75**, 53–58, doi:10.1175/1520-0493(1947)075<0053:ANOTMA>2.0.CO;2.

Simpson, R. H., 1947: Synoptic aspects of the intertropical convergence near Central and South America. *Bull Amer. Meteor. Soc.*, **28**, 335–346.

Simpson, R. H., 1948: On the slope of low-pressure axes as a criterion for deepening in the tropics. *Bull Amer. Meteor. Soc.*, **29**, 9–15.

Simpson, R. H., 1952: Evolution of the Kona storm, a subtropical cyclone. *J. Meteor.*, **9**, 24–35, doi:10.1175/1520-0469(1952)009<0024:EOTKSA>2.0.CO;2.

Simpson, R. H., 1952: Exploring eye of Typhoon Marge 1951. *Bull. Amer. Meteor. Soc.*, **33**, 286–298.

Simpson, R. H., 1954: Hurricanes. *Sci. Amer.*, **190**, 6.

Simpson, R. H., 1954: Structure of an immature hurricane. *Bull. Amer. Meteor. Soc.*, **35**, 335–350.

Simpson, R. H., 1959: Asymmetries in the hurricane. *Meteorology*, C. Ratanarat et al. Eds., Vol. 13, *Proceedings of the Ninth Pacific Science Congress of the Pacific Science Association*, Secretariat, Ninth Pacific Science Congress, 213–217.

Simpson, R. H., 1959: Hurricane cloudforms surveyed by reconnaissance aircraft. *Meteorology*, C. Ratanarat et al. Eds., Vol. 13, *Proceedings of the Ninth Pacific Science Congress of the Pacific Science Association*, Secretariat, Ninth Pacific Science Congress, 218–219.

Simpson, R. H., 1959: The West Indies Rawinsonde Network: A laboratory for problems in analysis and prediction of tropical weather. *Meteorology*, C. Ratanarat et al. Eds., Vol. 13, *Proceedings of the Ninth Pacific Science Congress of the Pacific Science Association*, Secretariat, Ninth Pacific Science Congress, 182–189.

Simpson, R. H., 1963: Liquid water in squall lines and hurricanes at air temperatures lower than $-40°C$. *Mon. Wea. Rev.*, **91**, 687–693, doi:10.1175/1520-0493(1963)091<0687:LWISLA>2.3.CO;2.

Simpson, R. H., 1965: Project STORMFURY—An experiment in hurricane weather modification. *Geofis. Int.*, **5**, 63–70.

Simpson, R. H., 1969: Understanding ocean weather. *Proc. Oceanology Int. 69,* Brighton, United Kingdom, Society for Underwater Technology.

Simpson, R. H., 1971 Mean-layer analyses for tracking and prediction of disturbances in the tropics. Preprints, *Symp. on Tropical Meteorology,* Honolulu, Hawaii, Amer. Meteor. Soc.

Simpson, R. H., 1973: Evacuations: Horizontal versus vertical. *Nation's Cities,* May, 45–46.

Simpson, R. H., 1973: Hurricane prediction: Progress and problem areas. *Science,* **181,** 899–907, doi:10.1126/science.181.4103.899.

Simpson, R. H., 1974: Hurricane prediction skill: Progress and prospects. Preprints, *Int. Tropical Meteorology Meeting,* Amer. Meteor. Soc., 145–150.

Simpson, R. H., 1974: Pilot plan for 'canes. *Museum,* June, 46–52.

Simpson, R. H., 1974: The complex killer. *Oceanus,* **17,** 22–27.

Simpson, R. H., 1975: Natural hazards. *Bull. Amer. Meteor. Soc.,* **56,** 478.

Simpson, R. H., 1975: On the design and evaluation of tropical cyclone seeding experiments. *Typhoon Modification: Proceedings of the WMO Technical Conference,* WMO 408, Secretariat of the World Meteorological Organization, 121–132.

Simpson, R. H., 1976: Tropical cyclone warning systems and their impact on industrial decision making. *Proceedings of the Symposium on the Impact of Tropical Cyclones on Oil and Mineral Development in North-west Australia,* Australian Government Public Service, 429–454.

Simpson, R. H., 1978: Hurricane prediction. *Geophysical Predictions,* Studies in Geophysics, National Academy of Sciences.

Simpson, R. H., 1978: On the computation of equivalent potential temperature. *Mon. Wea. Rev.,* **106,** 124–130, doi:10.1175/1520-0493(1978)106<0124:OTCOEP>2.0.CO;2.

Simpson, R. H., 1979: Impact of tropical cyclone winds. *Handbook of Abstracts: International Conference on Tropical Cyclones,* International Conference on Tropical Cyclones.

Simpson, R. H., 1980: Vertical evacuation: A viable alternative? *Proc. Third Int. Hurricane Conf.,* New Orleans, LA.

Simpson, R. H., 1981: Changes in the monsoon circulation of the South China Sea imposed by the moderate surge of 10-12 December 1978. *Proc. Int. Conf. on Early Results of FGGE and Large-Scale Aspects of Its Monsoon Experiments,* Tallahassee, FL, WMO.

Simpson, R. H., 1982: A hurricane hazard analysis for Longboat Key, Florida. Simpson Weather Associates, Inc., 54 pp.

Simpson, R. H., 1984: A risk analysis and preparedness decision system for use by coastal communities. Preprints, *15th Conf. on Hurricanes and Tropical Meteorology,* Miami, FL, Amer. Meteor. Soc.

Simpson, R. H., and L. G. Starrett, 1955: Further studies of hurricanes by aircraft reconnaissance. *Bull. Amer. Meteor. Soc.,* **36,** 459–468.

Simpson, R. H., and J. S. Malkus, 1963: An experiment in hurricane modification: Preliminary results. *Science,* **142,** 498, doi:10.1126/science.142.3591.498.

Simpson, R. H., and J. S. Malkus, 1964: Experiments in hurricane modification. *Sci. Amer.,* **211,** 27–37, doi:10.1038/scientificamerican1264-27.

Simpson, R. H. and Simpson, J., 1966: Why experiment on tropical hurricanes? *Trans. N.Y. Acad. Sci.,* **28,** 1045–1062, doi:10.1111/j.2164-0947.1966.tb02407.x.

Simpson, R. H., and J. M. Pelissier, 1971: Atlantic hurricane season of 1970. *Mon. Wea. Rev.,* **99,** 269–277, doi:10.1175/1520-0493(1971)099<0269:AHSO>2.3.CO;2.

Simpson, R. H., and J. R. Hope, 1972: Atlantic hurricane season of 1971. *Mon. Wea. Rev.,* **100,** 265–275, doi:10.1175/1520-0493(1972)100<0256:AHSO>2.3.CO;2.

Simpson, R. H., and P. J. Hebert, 1973: Atlantic hurricane season of 1972. *Mon. Wea. Rev.,* **101,** 323–333, doi:10.1175/1520-0493(1973)101<0323:AHSO>2.3.CO;2.

Simpson, R. H., and J. Simpson, 1975: Implications from the GATE dropwindsonde program regarding A-scale circulations. Preliminary scientific results of the GARP Atlantic Tropical Experiment, GATE Rep. 14, Vol. I, WMO and International Council of Scientific Unions, 1–11.

Simpson, R. H., and R. A. Pielke, 1976: Hurricane development and movement. *Appl. Mech. Rev.,* **29,** 601–609.

Simpson, R. H., and H. Riehl, 1981: *The Hurricane and Its Impact.* Louisiana State University Press, 398 pp.

Simpson, R. H., A. W. Johnson, and R. C. Gentry, 1956: Operation of the National Hurricane Research Project. *Weatherwise,* **9,** 111–119, doi:10.1080/00431672.1956.9927215.

Simpson, R. H., N. Frank, D. Shideler, and H. M. Johnson, 1968: Atlantic tropical disturbances, 1967. *Mon. Wea. Rev.,* **96,** 251–259, doi:10.1175/1520-0493(1968)096<0251:ATD>2.0.CO;2.

Simpson, R. H., N. Frank, D. Shideler, and H. M. Johnson, 1969: Atlantic tropical disturbances of 1968. *Mon. Wea. Rev.,* **97,** 240–255, doi:10.1175/1520-0493(1969)097<0240:ATDO>2.3.CO;2.

Simpson, R. H., A. I. Sugg, and Staff, 1970: The Atlantic hurricane season of 1969. *Mon. Wea. Rev.,* **98,** 293–306, doi:10.1175/1520-0493-98.4.293.

Simpson, R. H., P. K. Govind, and R. L. Holle, 1975: The GATE dropwindsonde program. *Bull. Amer. Meteor. Soc.,* **56,** 984–987.

Simpson, R. H., A. Miller, J. Simpson, and B. Fagan, 1979: TYMOD: Typhoon moderation. Final Rep. for PAGASA, Virginia Technology Inc., 128 pp.

Simpson, R. H., C. Warner, B. J. Morrison, and J. Simpson, 1981: Structure and dynamical milieu of monsoon eddies in the South China Sea and the related penetrative convection: The case of 16-17 December 1978. *Condensed Papers and Meeting Report: International Conference on Early Results of FGGE and Large-Scale Aspects of Its Monsoon Experiments,* Tallahassee, FL, WMO.

Simpson, R. H., B. Hayden, M. Garstang, and H. L. Massie, 1985: Timing of hurricane emergency actions. *Environ. Manage., 9,* 61–69, doi:10.1007/BF01871445.

Simpson, R. H., G. Holland, G., J. Simpson, 1989: Hurricane. *McGraw-Hill Encyclopedia of Science and Technology,* 6th ed. McGraw-Hill.

Wise, C. W., and R. H. Simpson, 1971: The tropical analysis program of the National Hurricane Center. *Weatherwise, 24,* 164–173, doi:10.1080/00431672.1971.9931538.

Technical Memoranda, Informal Reports, and Presentations

1955:   The hurricane fang and its importance to the middle Atlantic and New England states. *Woods Hole Oceanographic Associates Meeting,* New York, NY. (R. H. Simpson.)

1955:   On the structure of tropical cyclones as studied by aircraft reconnaissance. *Proceedings of the UNESCO Symposium on Typhoons,* Japanese National Commission for UNESCO, 129–150. [Available online at http://unesdoc.unesco.org/images/0013/001316/131628EB. pdf.] (R. H. Simpson.)

1955:   The Weather Bureau sizes up the hurricane problem. *Bell Teleph. Mag., 34,* 145–152. (R. H. Simpson.)

1956:   Objectives and basic design of the National Hurricane Research Project. National Hurricane Research Project Rep. 1, Dept. of Commerce, 6 pp. [Available online at http://www.aoml.noaa.gov/general/lib/TM/NHRP_01_1956.pdf.] (R. H. Simpson, N. E. LaSeur, R. C. Gentry, L. F. Hubert, and C. L. Jordan.)

1956:   Some aspects of tropical cyclone structure. *Proceedings of the Tropical Cyclone Symposium,* Bureau of Meteorology, 139–157. (R. H. Simpson.)

1957:   Hurricanes. Royal Canadian Institute (meeting). (R. H. Simpson.)

1958:   Mid-tropospheric ventilation as a constraint on hurricane development and maintenance. Preprints, *Technical Conf. on Hurricanes,* Miami Beach, FL, Amer. Meteor. Soc., D4-1–D4-10. (R. H. Simpson and R. Riehl.)

1958:   Details of circulation in the high-energy core of Hurricane Carrie. National Hurricane Research Project Rep. 24, Dept. of Commerce, 15 pp. [Available online at http://www.aoml.noaa.gov/general/lib/TM/NHRP_24_1958.pdf.] (National Hurricane Research Project Staff.)

1962: On the dynamics of disturbed circulation in the lower mesosphere. National Hurricane Research Project Rep. 57, Dept. of Commerce, 71 pp. [Available online at http://www.aoml.noaa.gov/general/lib/TM/NHRP_57_1962.pdf.] (R. H. Simpson.)

1962: A cloud seeding experiment in Hurricane Esther, 1961. National Hurricane Research Project Rep. 60, Dept. of Commerce, 30 pp. [Available online at http://www.aoml.noaa.gov/general/lib/TM/NHRP_60_1962.pdf. (R. H. Simpson, M. R. Ahrens, R. D. Decker.)

1963: Measurement of heat and momentum flux at the air/sea interface. *Third U.S.–Asian Military Weather Symp.*, John Hay Airbase, Philippine Islands.

1962: The unique development of the severe Atlantic coastal storm of March, 1962. *211th National Meeting of the Amer. Meteor. Soc.*, New York, NY, Amer. Meteor. Soc. (R. H. Simpson and J. Simpson.)

1964: *Hurricane Modification: Progress and Prospects.* University of California, 108 pp. (R. H. Simpson and J. S. Malkus.)

1967: Search for more adequate tools to describe circulation in tropical and equatorial latitudes. Naval Postgraduate School (meeting). (R. H. Simpson.)

1968: Role of the air/sea interface in the growth of hurricanes. *Symp. on Investigations and Resources of the Caribbean Sea and Adjacent Regions,* Curacao, British West Indies, WMO, UNESCO, and FAO. (R. H. Simpson.)

1969: Curbing hurricanes—The chances. (Interview) *U.S. News & World Report,* 1 September, 34–36. (R. H. Simpson.)

1969: Disturbances in the tropical and equatorial Atlantic. ESSA Tech. Memo. WBTM SR 47, PB 184 740. (R. H. Simpson.)

1969: A reassessment of the hurricane risk. National Red Cross Conf., Atlanta, GA.

1969: Synoptic analysis models for the tropics. Meteorological resources and capabilities in the '70's: Proceedings of the 5th AWS Meteorological Technical Exchange Conference, Air Weather Service Tech. Rep. 217, 189–200. [Available online at http://www.dtic.mil/get-tr-doc/pdf?AD=AD0862101.] (R. H. Simpson.)

1969: Your risk in hurricanes. Corpus Christi Rotary Club (meeting), May 1970.

1969: Don Franklin on—Weather. (Interview) Florida Afloat, 14–15 April.

1969: The hurricane as a machine of destruction. *Hurricane Foresight Meeting,* New Orleans, LA.

1969: Hurricane peak: Mid-September. (Interview) *U.S. News & World Report,* 10 August, 16.

1969: A reassessment of the hurricane prediction problem. ESSA Tech. Memo. WBTM SR 50, PB 189 846. (R. H. Simpson.)

1970: The Satellite Applications Section of the National Hurricane Center. ESSA Tech. Memo. WBTM SR 51, COM 71 00005. (R. H. Simpson and D. C. Gaby.)

1970: Tropical storms: Hurricanes. Nebraska Educational Television Council for Higher Education, Lincoln, NE.

1971: Atlantic hurricane frequencies along the U.S. coastline. NOAA Tech. Memo. NWS TM SR-58, 14 pp. [Available online at http://docs.lib.noaa.gov/noaa_documents/NWS/NWS_SR/TM_NWS_SR_58.pdf.] (R. H. Simpson and M. B. Lawrence.)

1971: The decision process in hurricane forecasting. NOAA Tech. Memo. NWS SR 53, COM 71 00336, 35 pp. (R. H. Simpson.)

1971: The hurricane: Perennial but increasing threat to our coast. Council of State Governments, Atlanta, GA (meeting). (R. H. Simpson.)

1971: Hurricane, yes or no. (Interview) NOAA, 1, 12–21. [Available online at http://docs.lib.noaa.gov/rescue/journals/noaa/QC851U461971jul.pdf.]

1971: Review of "Forecaster's Guide to Tropical Meteorology" (G. D. Atkinson, 225 pp.). *Eos, Trans. Amer. Geophys. Union*, **52,** 708–709.

1972: The hurricane risk and its potential impact. *J. Miami Academy of Sci.,* May, 1–12. (R. H. Simpson.)

1972: Hurricane vulnerability. Rotary Club of Miami (meeting). (R. H. Simpson.)

1972: Hurricane vulnerability. *New Orleans Press Association Meeting.* (R. H. Simpson.)

1973: Evacuation of coastal residents during hurricanes: A pilot study for Dade County, Florida; A report. Miami Federal Executive Board's Hurricane Shelters Committee, 40 pp. (R. H. Simpson, Chairman.)

1973: If hurricanes hit. (Interview) *U.S. News & World Report,* 14 August, 32–35.

1973: A decision procedure for application in predicting the landfall of hurricanes. NOAA Tech. Memo. NWS SR 71, COM 73-11663/AS. (R. H. Simpson and B. R. Jarvinen.)

1973: The neutercane: Small hybrid cyclone. Preprints, *Eighth Technical Hurricane Conf.,* Key Biscayne, FL, Amer. Meteor. Soc. (R. H. Simpson.)

1975: Assessing the impact of hurricanes on coastal structures. *Coastal Engineering Conf.,* Houston, TX, Amer. Meteor. Soc. (R. H. Simpson.)

1975: The GATE dropwindsonde program. Preprints, *Ninth Conf. on Hurricanes and Tropical Meteorology,* Key Biscayne, FL, Amer. Meteor. Soc. (R. H. Simpson and P. K. Govind.)

1976: A circulation analysis system for GATE area tropical disturbances. Preprints, *10th Conf. on Hurricanes and Tropical Meteorology,* Charlottesville, VA, Amer. Meteor. Soc. (R. H. Simpson, J. Cunningham, M. Zimner, and E. Hill.)

1976: Coastal hazard potentials. *First Conf. on Coastal Meteorology,* Virginia Beach, VA, 16–19. (R. H. Simpson and J. C. Freeman.)

1976: Transitions in African disturbances over the eastern Atlantic Ocean. Preprints, *10th Conf. on Hurricanes and Tropical Meteor.,* Charlottesville, VA, Amer. Meteor. Soc. (R. H. Simpson.)

1977:   Synoptic-scale disturbances in the eastern Atlantic Ocean. Preprints, *11th Technical Conf. on Hurricanes and Tropical Meteorology*, Miami Beach, FL, Amer. Meteor. Soc. (R. H. Simpson.).

1979:   Monitoring of the ocean–atmosphere environment to detect, understand, and predict hazardous weather and seas. Paper prepared for Computer Sciences Corporation, Bay St. Louis, MS. (R. H. Simpson and W. Frank.)

1979:   Physical nature and disaster potential of the hurricane. Lecture delivered at The Pennsylvania State University.

1979:   Will coastal residents reach safe shelter in time? Preprints, *13th Conf. on Hurricanes and Tropical Meteor.*, Orlando, FL, Amer. Meteor. Soc. (R. H. Simpson.)

1980:   Book review of *Disasters and the Mass Media. Bull. Amer. Meteor. Soc.* (R. H. Simpson.)

1980:   Implementation phase of the National Hurricane Research Project 1955–1956. Preprints, *13th Technical Conf. on Hurricanes and Tropical Meteor.*, Miami, FL, Amer. Meteor. Soc., 1–5. (R. H. Simpson.)

1980:   The quality of progress in predicting extreme events. *Potomac Geophysical Society Meeting.* (R. H. Simpson.)